Springer Proceedings in Complexity

Springer Proceedings in Complexity publishes proceedings from scholarly meetings on all topics relating to the interdisciplinary studies of complex systems science. Springer welcomes book ideas from authors. The series is indexed in Scopus.

Proposals must include the following:

– name, place and date of the scientific meeting
– a link to the committees (local organization, international advisors etc.)
– scientific description of the meeting
– list of invited/plenary speakers
– an estimate of the planned proceedings book parameters (number of pages/articles, requested number of bulk copies, submission deadline)

Submit your proposals to: christoph.baumann@springer.com

More information about this series at http://www.springer.com/series/11637

Aleksandra Przegalinska ·
Francesca Grippa · Peter A. Gloor
Editors

Digital Transformation of Collaboration

Proceedings of the 9th International COINs Conference

 Springer

Editors
Aleksandra Przegalinska
Kozminski University
Warsaw, Poland

Francesca Grippa
Northeastern University
Boston, MA, USA

Peter A. Gloor
MIT Center for Collective Intelligence
Cambridge, MA, USA

ISSN 2213-8684   ISSN 2213-8692   (electronic)
Springer Proceedings in Complexity
ISBN 978-3-030-48995-3   ISBN 978-3-030-48993-9   (eBook)
https://doi.org/10.1007/978-3-030-48993-9

This Springer imprint is published by the registered company Springer Nature Switzerland AG
The registered company address is: Gewerbestrasse 11, 6330 Cham, Switzerland

Preface

The 9th edition of the Collaborative Innovation Networks (COINs) conference was focused on "Digital Transformation of Collaboration" and represented a forum for researchers and practitioners from both academia and industry to meet and share cutting-edge advancements in the field of Social Network Analysis, Online Social Networks, Network Dynamics, and Collective Action. These topics are investigated with the usage of diverse, multidisciplinary methods, including data mining, statistics, machine learning, and deep learning.

The focal point of every Collaborative Innovation Networks conference is COINs—cyber teams of self-motivated people who collaborate by sharing ideas and information. COINs are powered by swarm creativity that enables a fluid creation and exchange of ideas, and is very often facilitated through networked technologies. This technological component of collaboration became prominent during this year's conference discussions and found its place among the contributions presented in this volume.

Throughout Collaborative Innovation Networks conferences, we have been looking at patterns of collaborative innovation and action. These frequently follow a similar path, from creator to COIN to Collaborative Learning Network (CLN), and finally to Collaborative Interest Network (CIN). This year, the focus on collaboration allowed us to look at different perspectives on collective action, not only between humans, but also between humans and technologies.

The contributions presented in this volume are organized according to four general themes: "body sensors", "emotions and morality", "human–machine interaction", and "interdisciplinary methods of collaboration". These are high-level classifications with clear overlap among different papers as the common denominator is the interplay of technology and humans in promoting collaboration and improved outcomes.

Within the theme of "body sensors", the authors focused on the abundance of information coming from human physiological signals. The authors of *"No Pain No Gain":Predicting Creativity Through Body Signals* presented interesting findings in predicting creativity through body signals. Their paper is introducing a novel methodology for finding creative individuals. They have assessed an individual's creativity with the Torrance Tests of Creative Thinking, and tracked the body signals with the sensors of a smartwatch measuring, among others, heart rate, acceleration, vector magnitude count, and loudness. Interestingly, these variables were combined with contextual data such as the environmental light level.

Using Body Signals and Facial Expressions to Study the Norms that Drive Creative Collaboration is focusing on norms that guide creative collaboration. The authors brought together concepts from complex systems theory and approach to cognition, to expose the dynamic nature of norms. The contribution *Measuring Audience and Actor Emotions at a Theater Play through Automatic Emotion Recognition from Face, Speech, and Body Sensors* describes a preliminary experiment to track the emotions of actors and audience in a theater play in Zurich through machine learning and AI. Eight actors were equipped with body sensing smartwatches. At the same time, the emotions of the audience were tracked anonymously using facial emotion tracking. The paper exposes an automated and privacy-respecting system to measure both audience and actor satisfaction during a performance.

Exploring the Impact of Environmental and Human Factors on Operational Performance of a Logistics Hub aimed at exploring environmental and human factors affecting the productivity of warehouse operators in material handling activities. The study was carried out in a semiautomated logistic hub and the data was collected using wearable sensors able to detect heart rate and human interactions. Yet another paper looking at how to measure performance and workload through data collected by wearables is *Measuring Workload and Performance of Surgeons Using Body Sensors of Smartwatches*. The authors' goal was to present introductory steps toward building an intelligent system to measure the workload and surgical performance of minimally invasive surgeons. Medical specialists who took part in the study wore a smartwatch with the Happimeter application that recorded a set of physiological and motion parameters during the surgical execution.

In *Exploring the Impact of Environmental and Human Factors on Operational Performance of a Logistics Hub*, the authors explored both human and environmental factors affecting the productivity of warehouse operators. Again, the data has been collected using wearable sensors able to detect human-related variables such as heart rate and human interactions, based on a smartwatch combined with a mobile application developed by the CCI MIT.

In the paper *Measuring Moral Values with Smartwatch-based Body Sensors*, the authors tackle the issue of predicting the moral values of individuals through their body movements measured with the sensors of a smartwatch. The personal moral values were assessed using the Schwartz Value Theory and the Big Five Personality Traits (OCEAN), and the data for all variables were gathered through the Happimeter app. Through multilevel mixed-effects generalized linear models, the results showed that sensor and mood factors can predict a person's values.

The main topics explored in *Heart Beats Brain—Measuring Moral Beliefs Through E-Mail Analysis* are linked to the previous paper exploring the issues of morality and individual behavior. This paper investigates the very possibility of automatic measurement of moral values through hidden "honest" signals in the person's textual communication. The authors measured the e-mail behavior of 26 users through their e-mail exchanges, calculating their seven "honest signals of collaboration" (strong leadership, balanced contribution, rotating leadership, responsiveness, honest sentiment, shared context, and social capital). According to the authors, these honest signals explained 70 percent of their moral values measured with the moral foundations survey.

Similarly, *Identifying Virtual Tribes by Their Language in Enterprise Email Archives* explores automatic grouping of employees into virtual tribes based on their language and values. The authors used the *Tribefinder* tool to analyze two email archives, the individual mailbox of an active academic and corporate consultant, and the Enron email archive. In *the political debate on immigration in the election campaigns in Europe*, the authors' study explored the political debate on immigration during the election campaigns of France and Italy over the last 3 years. They perform Emotional Text Mining with the aim of identifying the sentiment surrounding immigration, and how immigrants are portrayed, in the online Twitter debate.

Text mining and SNA play a major role in *Brand Intelligence Analytics*. The authors describe the functionalities of the SBS Brand Intelligence App (SBS BI), designed to assess brand importance and provide brand analytics through the analysis of textual data. They use a case study focused on the 2020 US Democratic Presidential Primaries.

Finally, contributions in this section explored data-driven methods to understand how ideologies and religions interact with each other. In *Finding patterns between religions and emotions*, the main question was whether specific religions and emotions are connected. Based on Twitter data, authors trained the model to create "tribes" for four main religions and four basic emotions. Similarities and differences between tribes were analyzed using the content of the tweets. Applying similar methodology, in *Virtual Tribes: Analyzing Attitudes towards the LGBT Movement* the authors investigated the application of machine learning techniques that allow conclusions from users' behavior and used language on Twitter concerning their attitudes toward the LGBT movement. By using an adjusted procedure of the Cross-Industry Standard Process for Data Mining, the authors created a prediction model and provided instructions for its deployment.

The next broad theme discussed in this volume is related to collective intelligence meeting artificial intelligence investigated in the context of human–machine interaction. In *Digital Coworker-Human-AI Collaboration in Work Environment*, collaboration is investigated from the perspective of the relation between humans and virtual assistants. Contrary to the mainstream opinion in the general public that automation will hold negative societal implications, such as job loss, the authors argue that broad introduction of AI-based tools for knowledge professions will lead to job effectiveness and increased job satisfaction.

In *Collaborative Innovation Network in Robotics*, COINs were explored among robots. Here again, human–machine and machine–machine interaction is researched in the context of collaboration rather than competition. The study employed Knowledge Building pedagogy and technology to integrate robotics into subjects like mathematics.

Fantastic Interfaces and Where to Regulate Them is an interdisciplinary work, bridging Law, Social Sciences, and Human–Computer Interaction (HCI) design and focused on speech interfaces. This work address three central aspects from a legal perspective: (i) the possibility to lie; (ii) the possibility to breach the law; (iii) the ability to interpret an order. It provides an overview of the correct hermeneutical approach to frame legal paradigms, highlighting the key legal aspects to consider when analyzing the phenomenon of "interactive artificial agents".

Finally, the interdisciplinary methods theme is opened by *A structured approach to GDPR compliance*, focused on analyzing legal frameworks and regulations. The authors recognized that the General Data Protection Regulation framework (GDPR) had profoundly changed the legislative approach to the protection of personal data by the European Union and that current organizations must adopt a proactive approach based on accountability. In their paper, they proposed a structured approach, based on business process modeling, to support compliance with the GDPR.

The authors of *Mapping Design Anthropology: Tracking the Development of an Emerging Transdisciplinary Fields* applied social network analysis to investigate the human and nonhuman actors (i.e., people and institutions) that have contributed to design anthropological practice and theories.

Combining Social Capital and Geospatial Analysis to measure the Boston's Opioid Epidemic is an academic paper that combines theoretical contribution to practical recommendations to solve a public health problem. The authors recognized that traditional drug treatment interventions have mainly focused on the individuals without taking into account their environment and existing social support networks. By combining a social capital framework with geospatial research, the authors managed to map opioid resilience of a community by focusing on the Boston neighborhoods. Their analysis clearly shows that in neighborhoods with an active community, the risk of opioid-related deaths is lower.

In *Reward-Based Crowdfunding as a Tool to Constitute and Develop Collaborative Innovation Networks*, the authors argue that reward-based crowdfunding provides a practicable tool to constitute and develop COINs. In their paper, they develop a conceptual framework of this form of crowdfunding that could be applied to support the constitution and development of COINs.

In *An Ecosystem for Collaborative Pattern Language Acquisition*, the authors describe an ecosystem for acquiring pattern languages from the perspectives of constructivism and collaborative way and introduce a web system called "Presen Box" that assists in pattern languages acquisition according to the ecosystem.

In today's emerging networked world, technologies in general and AI or IoT, in particular, play a more and more important societal and economic role. Investigating fruitful collaboration and looking for the best practices is extremely important. Understanding crowd dynamics and collective action allows us to look for better pathways of collaboration in the future, both between humans themselves, and between humans and machines.

Warsaw, Poland   Aleksandra Przegalinska
Boston, USA   Francesca Grippa
Cambridge, USA   Peter A. Gloor

Acknowledgments

This volume combines selected contributions presented at the 9th International Conference on Collaborative Innovation Networks "Digital Transformation of Collaboration" held in Warsaw, Poland on October 8th and 9th, 2019. This was an occasion for international scholars and business practitioners to share their latest research findings and discuss opportunities to collaborate in the area of collaborative emerging technologies and the application of various methodologies, including—but not limited to—Social Network Analysis, data mining, AI, and deep learning.

We would like to thank the sponsors of the conference who helped bring together international scholars, researchers, and practitioners in Warsaw. These include Digital University together with Masters & Robots Conference, Microsoft, and Grupa Pracuj, as well as Kozminski University, Wayne State University, and the MIT Center for Collective Intelligence.

We are also grateful for the valuable insights provided by the Conference Program Committee members and to the many reviewers who provided both insightful feedback and constructive criticisms.

Contents

Contributors

Davide Aloini University of Pisa, Pisa, Italy

Keith April Araño Politecnico di Milano, Milan, Italy

Earlene Avalon Northeastern University, Boston, MA, USA

Michael Beier University of Applied Sciences of the Grisons, Swiss Institute for Entrepreneurship, Chur, Switzerland

Giulia Benvenuti University of Pisa, Pisa, Italy

Moritz Bittner Universität zu Köln, Cologne, Germany

José L. Campos Jesús Usón Minimally Invasive Surgery Centre, Cáceres, Spain

Antonio Capodieci Università del Salento, Lecce, Italy

Andrea Fronzetti Colladon Department of Engineering, University of Perugia, Perugia, Italy

David Dettmar Universität zu Köln, Cologne, Germany

Riccardo Dulmin University of Pisa, Pisa, Italy

J. A. Edelman HPI-Stanford Design Thinking Research Program, Hasso Plattner Institute, Potsdam, Germany

Joscha Eirich University of Bamberg, Bamberg, Germany

Jens Fehlner Otto-Friedrich-Universität Bamberg, Bamberg, Germany

Sonja Fischer Otto-Friedrich-Universität Bamberg, Bamberg, Germany

Sebastian Früh University of Applied Sciences of the Grisons, Swiss Institute for Entrepreneurship, Chur, Switzerland

S. K. Ghosh University of Potsdam, Potsdam, Germany

Peter A. Gloor MIT Center for Collective Intelligence, Cambridge, MA, USA; Massachusetts Institute of Technology, Cambridge, MA, USA

Francesca Greco Sapienza University of Rome, Rome, Italy

Francesca Grippa Northeastern University, Boston, MA, USA

Emanuele Guerrazzi University of Pisa, Largo Lucio Lazzarino, Pisa, Italy

Takashi Iba Faculty of Policy Management, Keio University, Tokyo, Japan

Yuki Kawabe Faculty of Environment and Information Studies, Keio University, Tokyo, Japan

Ahmad Khanlari University of Toronto, Toronto, ON, Canada

Annika Lurz Otto-Friedrich-Universität Bamberg, Bamberg, Germany

Luca Mainetti Università del Salento, Lecce, Italy

Alexandra Manger Otto-Friedrich-Universität Bamberg, Bamberg, Germany

Christine Miller Savannah College of Art and Design, Savannah, USA

Valeria Mininno University of Pisa, Pisa, Italy

Diego Morejon Jaramillo Universität Bamberg, Bamberg, Germany

Lee Morgan MIT Center for Collective Intelligence, Cambridge, MA, USA

R. Mukherjee University of Potsdam, Potsdam, Germany

B. Owoyele HPI-Stanford Design Thinking Research Program, Hasso Plattner Institute, Potsdam, Germany

Alessandro Polli Sapienza University of Rome, Rome, Italy

Aleksandra Przegalinska Kozminski University, Warsaw, Poland

Ken Riopelle Wayne State University, Detroit, USA

Gianluigi M. Riva School of Information and Communication Studies, University College Dublin, Dublin, Ireland; Massachusetts Institute of Technology, Media Lab, Cambridge, MA, USA

Cordula Robinson Northeastern University, Boston, MA, USA

Francisco M. Sánchez-Margallo Jesús Usón Minimally Invasive Surgery Centre, Cáceres, Spain

Juan A. Sánchez-Margallo Jesús Usón Minimally Invasive Surgery Centre, Cáceres, Spain

J. Santuber HPI-Stanford Design Thinking Research Program, Hasso Plattner Institute, Potsdam, Germany

Konrad Sowa Kozminski University, Warsaw, Poland

Alessandro Stefanini University of Pisa, Pisa, Italy

Marius Stein University of Cologne, Cologne, Germany

Lirong Sun MIT Center for Collective Intelligence, Cambridge, MA, USA; University of Chinese Academy of Sciences, Haidian District, Beijing, China

Maximilian Johannes Valta Universität Bamberg, Bamberg, Germany

Qi Wen Tsinghua University, Beijing, China

Michael Wood Northeastern University, Boston, MA, USA

Part I
Body Sensors and Big Data

Chapter 1
"No Pain No Gain": Predicting Creativity Through Body Signals

Lirong Sun, Peter A. Gloor, Marius Stein, Joscha Eirich, and Qi Wen

Abstract Creative people are highly valued in all parts of the society, be it companies, government, or private life. However, organizations struggle to identify their most creative members. Is there a "magic ingredient" that sets the most creative individuals of an organization apart from the rest of the population? This paper aims to shed light on a part of this puzzle by introducing a novel method based on analyzing body language measured with sensors. We assess an individual's creativity with the Torrance Tests of Creative Thinking, while their body signals are tracked with the sensors of a smartwatch measuring heart rate, acceleration, vector magnitude count, and loudness. These variables are complemented with external environmental features such as light level measured by the smartwatch. In addition, the smartwatch includes a custom-built app, the Happimeter, that allows users to do mood input in a two-dimensional framework consisting of pleasance and activation. Using multi-level regression, we find that people's creativity is predictable by their body sensor readings. We thus provide preliminary evidence that the body movement as well as environmental variables have a relationship with an individual's creativity. The results also highlight the influence of affective states on an individual's creativity.

L. Sun · P. A. Gloor (✉)
MIT Center for Collective Intelligence, 245 First Street, 02142 Cambridge, MA, USA
e-mail: pgloor@mit.edu

L. Sun
University of Chinese Academy of Sciences, Haidian District, 100190 Beijing, China

M. Stein
University of Cologne, Albertus Magnus Platz, 50923 Cologne, Germany

J. Eirich
University of Bamberg, Kapuzinerstr 16, 96047 Bamberg, Germany

Q. Wen
Tsinghua University, Haidian District, 100084 Beijing, China

© Springer Nature Switzerland AG 2020
A. Przegalinska et al. (eds.), *Digital Transformation of Collaboration*,
Springer Proceedings in Complexity,
https://doi.org/10.1007/978-3-030-48993-9_1

3

Introduction

Creative thinking is the foundation for art, science, and technology. It is at the core of societal advancement [1, 3], leading to products, ideas, or processes that are novel and useful [14]. However, what kind of features do highly creative individuals display that sets them apart from the rest of the population? Is there an automated way to distinguish them from others? Current best practices to identify the most creative individuals are based on surveys and cognitive assessments [19]. In this paper we investigate the research question if the body language of a person might also predict her/his creativity.

Based on the rapid growth of mobile computing and sensor technology, we introduce a novel way to predict creativity through data collected from a variety of wearable sensors. In this study, we aim to advance an integrative view of a person's creativity through the lens of body sensors, mood states, and external environment features. Using commercially available smartwatches, we built a body sensing system called "Happimeter" [6] that collects data in three broad categories for objectively measuring physical activity: body movement, physiology, and context information. We explore the associations between these factors and the five constructs of their creativity obtained from the Torrance Test of Creative Thinking: Fluency, Originality, Elaboration, Abstractness of Titles, and Resistance to Premature Closure [23, 25]. We also investigate how pleasance and activation levels [9] collected by the Happimeter app will influence the five aspects of a person's creativity.

Theoretical Background

The Measurement of Creativity

Creativity of a solution to a problem is characterized by assessing the novelty (e.g., solutions have less frequent features) and utility (i.e., solutions satisfy precise needs) of the solution [21]. A broad set of theories and models are discussed in the research literature on cognitive psychology [11], revealing different ways to measure an individual's creativity. In this paper, we use the Torrance Tests of Creative Thinking (TTCT). Researchers have employed the Torrance Tests of Creative Thinking (TTCT, [22, 23, 25]) for more than four decades, and this measure continues to dominate the field when it comes to the testing of individuals' creativity from kindergarten through adulthood.

The TTCT contains two parts, the verbal part and the figural part. It consists of five activities, some of which need to be answered by drawing, while others are answered in writing. As is common with such tests, it has to be completed under time pressure. The current form of the TTCT includes scores for Fluency, Originality, Elaboration, Abstractness of Titles, and Resistance to Premature Closure (see [10], for details).

Table 1.1 Five dimensions of creativity in TTCT

Aspects	Definition
Fluency (flu)	The number of relevant ideas; shows an ability to produce a number of figural images
Originality (origin)	The number of statistically infrequent ideas; shows an ability to produce uncommon or unique responses
Elaboration (elab)	The number of added ideas; demonstrates the subject's ability to develop and elaborate on ideas
Abstractness of titles (abs)	The degree beyond labeling; based on the idea that creativity requires an abstraction of thought
Resistance to premature closure (res)	The degree of psychological openness; based on the belief that creative behavior requires a person to consider a variety of information when processing information and to keep an "open mind"

The five subscales and information about scoring and the content measures are shown in Table 1.1.

Torrance [24] discouraged interpretation of scores as a total score or a static measure of a person's ability and warned that using a composite score might be misleading because each subscale score has independent meaning. Instead, Torrance encouraged the interpretation of subscale scores separately. We followed these instructions in our study by using the five constructs of creativity instead of a composite score. While some individuals were wearing the watch over extended periods of time, we restricted analysis of the sensor readings to a time window of three days around the time when the individuals took the test. Some participants only wore the watch for a few days around the time when taking the test, while others had been wearing it for years.

Factors Influencing Creativity

Prior studies reveal that factors which influence creativity can be broadly classified as either domain- and creativity-relevant factors or affect and external factors [7]. The Torrance Test of Creative Thinking aims to predict creative performance in general, outside of a given domain (for a discussion, see [18]). Thus, in this article, domain-relevant skills are not our focus. Also, Hennessey and Amabile [7] argue that there are factors on seven levels that will influence a person's creativity, ranging from the biological basis to affect/cognition/training, from the environment to culture and society. Here we focus on the biological, affective, and environmental differences between individuals to reveal their creativity by using data from sensor devices and answers from users.

Three general categories of sensors can be used for measuring parts of the above-mentioned information: movement sensors, physiological sensors, and contextual sensors [2]. Movement sensors can be used to measure human physical activities. Among these devices, accelerometers are currently the most widely used sensors for human physical activity monitoring. Physiologic sensors include measurement of heart rate, blood pressure, temperature (skin and core body), heat flux, and so on. To date, heart rate monitoring remains the most common sensor for physiologic monitoring and is used in our research. Contextual sensors are concerned with assessing the context of the environment in which the physical activity is being performed. Compared to movement and physiological sensors, contextual sensors are relatively new and have great potential to help describe the relationship between physical activity and various environmental features. We obtain the light information of the imminent environment by our "Happimeter" sensor system which combines smartwatches and smartphones. With regard to affective states, we implement our analysis based on the two dimensions pleasance and activation, according to the Circumplex Model of Affects [9]. We explored other models of emotion such as the six-dimensional model of Plutchik.

There is evidence that highly creative individuals have a tendency to be physiologically overactive [13]. Moreover, a large body of literature has investigated affective impact on creativity. For instance, Isen et al. [8] found that participants performed better on creative problem-solving tasks when they experienced positive affect than participants in a negative or neutral affective state. Thus, they argue that positive affect fosters creativity. Similarly, Murray et al. [16] also found that positive affect increased creativity. However, instead of highlighting that positive affective state is better than negative affect for fostering creativity, George and Jing [5] found that negative affect can help identify when a conscious effort is needed to refine and improve creative outcomes. This contradiction in scientific evidence about the relationships between affect and creativity is also addressed in our article. As for the influence of the external environment, we argue that features of the surroundings might modify the performance of generating novel and useful solutions to creative problems. We also investigated the influence of light, more specifically, the level of illumination. Some studies show that a darker environment is better for out-of-the-box thinking since bright lights give people the impression that they are under surveillance, and thus less free to take risks. But other scholars also posit that low-light conditions discourage the eye from focusing on details, leaving people free to get involved in abstract mental processing of creative thinking [19].

To the best of our knowledge, less work has investigated the utility of sensing devices for modeling creativity. For instance, Muldner and Burleson [15] applied machine learning to data from eye tracking, a skin conductance bracelet, and an EEG sensor to predict creativity. They found reliable differences in sensor features characterizing low versus highly creative students. As this is an emerging field, more work is needed to explore how to apply body sensors to creative problem-solving. Our research provides a step in this direction.

The theoretical framework is shown in Fig. 1.1.

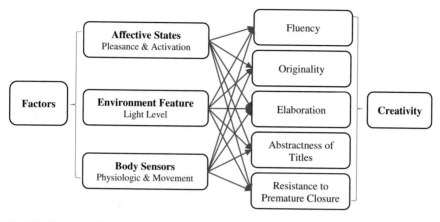

Fig. 1.1 Theoretical framework

Methodology

Participants

Participants were members of the 2017 Collaborative Innovation Networks (COINs) conference that took place in Detroit, USA from September 14 to 17, 2017, and of two student block seminars held in Bamberg and Cologne, Germany, from October 10 to 17, 2017. They completed the TTCT using paper and pencil. The participants wore smartwatches provided by our research group and downloaded the Happimeter app from the Google Play or iTunes store, and installed it on their smartphones and smartwatches. The analysis reported here includes all users who provided data on all dependent and independent variables required for this analysis. Some users had technical problems connecting the smartwatch to their phones, or to install the app, these users were excluded from the analysis.

A total of 50 users' creativity tests are collected (23 from the conference and 27 from the seminar) and their creativity score is graded by three raters according to our research design. To demonstrate consistency among observational ratings provided by these coders, the assessment of inter-rater reliability (IRR) is necessary for verifying the reliability of our data. IRR analysis aims to determine how much of the variance in the observed scores is due to variance in the true scores after the variance due to measurement error between coders has been removed. The results of IRR are shown in Table 1.2.

In addition, we obtained the creativity score from their TTCT results, and combined it with momentary self-reports of affective mood states and sensor data collected by the Happimeter. The Happimeter app trains mood prediction system by repeatedly asking two questions: [1] How pleasant do you feel? [2] How active do you feel? The user chooses his affective states levels with a scale of 0–2, where 0 represents low pleasance or activation, while 2 corresponds to high pleasance or

Table 1.2 Inter-rater reliability

Dimensions of creativity	IRRs for conference	IRRs for seminars
Fluency	0.983	0.956
Originality	0.852	0.790
Elaboration	0.957	0.841
Resistance	0.828	0.65
Abstractness	0.759	0.479

activation. After matching affective states data, sensor data, and TTCT scores and doing data filtering, we got a total of 8339 records with sensor data for 37 users, among whom 57% are male and 43% are female. Given that these users may or may not have provided their affect information at the same frequency as their sensor data, the analyses described in the results section uses an average level of pleasance and activation on each day to match with the sensor data records. The majority of the participants (62%) are from Europe, and 16% each from Asia and North America.

Model

Variables

The predictor measures are shown in Table 1.3. As some users did not report their age, we did not include it into the analysis. This is supported by the results of Lee and Kyung [12], who found that age was not an influential factor of creativity.

Multilevel Analysis

We use multilevel analysis with levels sensor and user. The variability in the outcome can be thought as being either within a user or between users. The sensor data level observations are not independent, which means that for a given user, sensor data records are related to each other. The multilevel mixed-effects generalized linear model, using the Stata mixed procedure, was performed with 8339 sensor data records (Level 1) across 37 individuals (Level 2) to control for the nested data structure. The independent variables pleasance and activation were collected through experience-based sampling [6] by polling users at random times per day on the smartwatch by asking them the questions "how active do you feel?" and "how pleasant do you feel?", and the user could then enter this information using a slider shown on the touchscreen of the watch.

Table 1.3 Variables

	Category	Variables	Definition
Dependent variables	Five aspects of creativity	Flu	The standardized score of *Fluency* derived from TTCT results
		Origin	The standardized score of *Originality* derived from TTCT results
		elab	The standardized score of *Elaboration* derived from TTCT results
		abs	The standardized score of *Abstractness of Titles* derived from TTCT results
		res	The standardized score of *Resistance to Premature Closure* derived from TTCT results
Independent variables	Affective states	pleasance	Self-reported scores for pleasance, range from 0 to 2
		Activation	Self-reported scores for activation, range from 0 to 2
	Physiologic sensors	avgbpm	The average number of heart beats per minute (standardized)
		varbpm	The variance of heart rate within a day (standardized)
	Contextual sensors	avglight	The light level of the environment (standardized)
		varlight	The variance of light level within a day (standardized)
	Movement sensors	VMC	The vector magnitude counts of the user (standardized)
		avgacc	The magnitude of the acceleration of the user's movement in the physical space (standardized)
		varacc	The variance of acceleration of the user's movement within a day (standardized)

(continued)

Table 1.3 (continued)

	Category	Variables	Definition
Control variables	User profile	Gender	Gender $= 1$, male, otherwise, female
	User location	Continent 1	Continent $1 = 1$ if the user is from Asian, otherwise, 0
		Continent 2	Continent $2 = 1$ if the user is from Europe, otherwise, 0
		Continent 3	continent $3 = 1$ if the user is from North America, otherwise, 0
		Continent 4	Continent $4 = 1$ if the user is from Oceania, otherwise, 0
		Continent 5	continent $5 = 1$ if the user is from South America, otherwise, 0

Results

Table 1.5 presents correlations for all explanatory variables along with descriptive statistics, which indicates that multicollinearity should not pose a problem. No value is higher than the threshold of 0.7 that is used as a rule of thumb for collinear relationships. We also checked the variation inflation factors (VIFs) and found that the VIFs for all principal variables are below the rule-of-thumb cutoff of 10 (the average of VIFs is 2.08), indicating no serious problem with multicollinearity [17]. Our last test was to check for heteroscedasticity in the data using the Breusch-Pagan test. Results indicate that heteroscedasticity is not a problem with the data.

With the null model (permitting random intercepts only), we calculate the intraclass correlations (ICCs) of all creativity dimensions, which are shown as Table 1.4. ICC is an indication of the extent to which sensor data of the same user are similar on their value scores relative to the total variation in sensor data of all users. An ICC value of 0 signifies complete observation independence within a user, while an ICC

Table 1.4 Interclass coefficients of null models

	ICC	Std. err.	95% conf.	Interval
Fluency	0.368	0.0762	0.234	0.525
Originality	0.31	0.0726	0.188	0.467
Elaboration	0.397	0.0783	0.257	0.555
Abstractness	0.442	0.0813	0.293	0.602
Resistance	0.464	0.0804	0.314	0.62

value of 1 indicates that differences in the outcome variable are completely dependent on the grouping variable. Therefore, ICC is used to verify if the nested design is suitable using a multilevel model. A review of the ICCs of our model shows that multilevel regression is a good method for our data and analysis since the ICCs of these creativity dimensions varies from 0.3 to 0.5, meaning that significant similarities exist in each group (user). For all dimensions, 30–50% of the creativity variance can be explained by the differences between users, while 50–70% can be accounted for by variables on the sensor data level.

Second, we test the predictive effects of body sensors, external environment feature, and affective states on creativity. We build on the previous model by adding random intercepts and fixed-effect predictors from level 1 or/and level 2 into our regressions. Table 1.6 provides regression results. Models 0–4 serve as baseline models that include only control variables (also level 2 variables). As shown in our sample, females tend to be more creative than men in all dimensions: Fluency, Originality, Elaboration, Abstractness of titles, and Resistance to premature closure. As for the continental difference, only taking continents with large samples into consideration, the conclusion can be made that Europeans (continent 2) are the most creative people, while Asian (continent 1) people are the least creative. The creativity scores for individuals from North America (continent 3) are in-between. As the sample size for the South American and Oceania participants is too small, they are not included in this analysis.

Models 5–9 test the predictive and influential power of body sensors, environmental feature, and affective states on five subscores of creativity. The results support that all three groups of variables are significantly related to creativity but vary in the degrees and directions of the influence they exert on.

The results show that sensor features can reliably distinguish high creativity individuals from the low creativity ones. The average of heartbeat (standardized heartbeat average) negatively influences fluency and elaboration subscores of creativity but is positively related to abstractness. The variance of heart rate within a day is mostly positively related to the five dimensions of creativity, with positive coefficients for fluency, originality, abstractness, and elaboration but negatively related to resistance. As for acceleration and VMC (vector magnitude counts), we also find partial support for their relationships with creativity.

Table 1.5 Descriptive statistics and correlations

		Mean	Std.Dev.	Min	Max	1	2	3	4	5	6	7	8	9	10
1	avgbpm	79.49	19.06	40	200	1.00									
2	varbpm	247	102.1	0	466.6	0.13**	1.00								
3	avglight	1.612	0.992	-1	4	0.00	-0.05**	1.00							
4	varlight	11.93	6.383	0	22.02	0.04**	0.10**	0.35**	1.00						
5	avgacc	888.8	138.3	22.98	1299	-0.04**	0.01	0.01	-0.05**	1.00					
6	varacc	282774	213845	15.07	1569000	0.06**	0.05**	0.05**	0.12**	-0.48**	1.00				
7	vmc	2202	2119	0	15938	0.29**	0.07**	-0.05**	-0.01	-0.19**	0.24**	1.00			
8	varvmc	30444	11035	0	47452	0.07**	0.67**	0.01	0.44**	-0.03*	0.11**	0.10**	1.00		
9	pleasance	1.36	0.443	0	2	-0.03*	-0.06**	0.12**	0.03**	0.03**	0.00	-0.04**	-0.19**	1.00	
10	activation	0.74	0.381	0	2	0.03**	-0.23**	0.04**	0.23**	-0.04**	0.07**	0.00	-0.08**	0.30**	1.00
11	gender	0.677	0.468	0	1	-0.07**	0.23**	-0.08**	-0.20**	0.04**	0.01	0.01	-0.05**	-0.08**	-0.10**

Note: ** $p < 0.01$, * $p < 0.05$

Note ** $p < 0.01$, * $p < 0.05$

Table 1.6 Results of multilevel regressions

Variables	Model0	Model1	Model2	Model3	Model4	Model5	Model6	Model7	Model8	Model9
	flu	origin	abs	elab	res	flu	origin	abs	elab	res
Gender	−11.13***	−11.06***	−11.47***	−3.102***	−14.42***					
Continent 1	−61.62***	−43.19***	−74.14***	−49.60***	−50.95***					
Continent 2	−46.91***	−29.42***	−39.02***	−38.81***	−30.40***					
Continent 3	−52.14***	−38.25***	−52.17***	−47.12***	−47.07***					
Continent 4	−22.45***	−16.16***	−39.26***	17.31***	7.371***					
o.continent5	–	–	–	–	–	–	–	–	–	–
Pleasance						−8.743***	−3.598***	−6.113***	−13.34***	−10.49***
Activation						5.977***	−1.762***	−5.542***	7.883***	0.664
stdavgbpm						−1.222***	−0.212	0.430***	−0.726***	−0.14
stdvarbpm						2.438***	2.297***	6.116***	4.161***	−2.873***
stdavglight						1.438***	1.638***	1.476***	1.096***	1.165***
stdvarlight						−5.255***	−4.697***	−2.933***	−5.451***	6.743***
stdavgacc						0.618***	0.530***	0.2	−0.00364	−0.0807
stdvaracc						−0.0959	0.0374	0.702***	0.0211	−0.305*
stdvmc						0.351**	0.274	−0.0919	0.324***	0.441***
stdvarvmc						−3.648***	−4.625***	0.810***	−2.268***	0.163
Constant	151.8***	136.6***	148.3***	136.4***	139.8***	156.2***	136.1***	150.1***	144.9***	166.9***
Observations	8.339	8.339	8.339	8.339	8.339	8.339	8.339	8.339	8.339	8.339
Number of groups	37	37	37	37	37	37	37	37	37	37

Note *** $p < 0.01$, ** $p < 0.05$, * $p < 0.1$

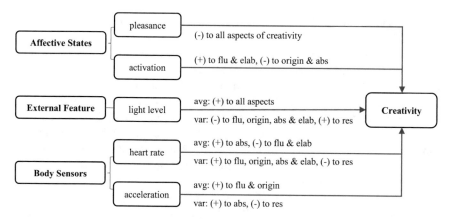

Fig. 1.2 Significant predictors of creativity

Another important finding is that the external light level has great impact on creativity (both the average light level and variance of light level within a day). While increasing light level will enhance all subscores of creativity, the variance of light will lead to the reduction of four dimensions of creativity (fluency, originality, elaboration, abstractness).

Moving on to the influence of affective states, both the scores of pleasance and activation play a prominent role regarding creativity. One interesting finding is that pleasance would actually impede a person's creativity. Our results also provide evidence for the relationship between activation and creativity, while activation promotes fluency and elaboration, and it has adverse effects on originality and abstractness.

Figure 1.2 summarizes our results, with (+) referring to a positive relationship, while (−) indicates a negative one.

Discussion

Literature focusing on creative thinking can be divided into research about the creative product, process, person, and environment [4]. This study focused on creative thinking skills of a person, investigating the influence of body sensors, environmental features, and affective states. We find strong contributors from the biological, psychological, and environmental sides. Comparing the specific indexes, we find that while both the average and variance of heart rate, light level, and pleasance and activation have relatively stronger influences on creativity, accelerometer data (VMC and acceleration) has relatively small, almost negligible, effect size. We independently investigated the five aspects of creativity since the TTCT is an instrument used to measure different constructs and prior literature advised them being explored separately. Our results show distinct differences in the impact of body sensor, mental states, and environmental features on difference subscores of the creativity measured by TTCT.

Also, on the individual level, similar to Lee and Kyung [12], we find gender difference in creativity, which reveals that females tend to score higher in different constructs than men. Moreover, the cultural differences revealed by Saeki et al. [20], which assessed cross-cultural creativity differences between American and Japanese students, are confirmed in our study. We identify regional inequalities of creativity between study participants from different continents, and find that Europeans are most creative, followed by North Americans, and Asians. This illustrates that creativity cannot be isolated from the socio-cultural context in which an individual lives and works.

This study contributes to literature about creativity by both supporting previous findings and providing new insights. We provide evidence for the gender and region difference in creativity. By new insights, we refer to the introduction of a novel method to measure individuals' creativity based on analyzing body language, environmental feature, and mood states. We also demonstrate how technology, specifically sensor-based systems like the Happimeter, might be used to collect personality characteristics in a non-intrusive way, without the need to fill out surveys. However, this study also has some limitations. First, the dataset is quite small, with 37 participants. Future research is needed to replicate the findings related to the five constructs of creativity with a larger size group of people. Second, this study only pays attention to a set of limited variables related to body sensor, affective states, and environmental feature. As there is rapid development in this field, it is desirable to collect in future research data from different types of sensors. For instance, researchers could integrate latest results on analyzing stress levels by adding more psychology-related variables, or more variables describing types of noise in the environment. This research did not explore group problem-solving or an actual creative solution. Rather it was a correlational analysis of independent variables with a standardized creativity assessment. Future research may explore creative teams on real-world problems. With the current work we have barely scratched the surface of this exciting new area of research.

All procedures performed in studies involving human participants were in accordance with the ethical standards of the institutional and/or national research committee with the Helsinki declaration and its later amendments or comparable with ethical standards.

References

1. T.M. Amabile, Creativity and innovation in organizations (1996)
2. K.Y. Chen, K.F. Janz, W. Zhu, R.J. Brychta, Re-defining the roles of sensors in objective physical activity monitoring. Med. Sci. Sports Exerc. **44**(1 Suppl 1), S13 (2012)
3. M. Csikszentmihalyi, *The Systems Model of Creativity: The Collected Works of Mihaly Csikszentmihalyi*. Springer, (2015)
4. C.R. Friedel, R.D. Rudd, Creative thinking and learning styles in undergraduate agriculture students. J. Agric. Educ. **47**(4), 102 (2006)

5. J.M. George, Z. Jing, Understanding when bad moods foster creativity and good ones don't: the role of context and clarity of feelings. J. Appl. Psychol. **87**(4), 687–697 (2002)
6. P.A. Gloor, A.F. Colladon, F. Grippa, P. Budner, J. Eirich, Aristotle said "happiness is a state of activity"—Predicting Mood Through Body Sensing with Smartwatches. J. Sys. Sci. Syst. Eng. **27**(5), 586–612 (2018)
7. B.A. Hennessey, T.M. Amabile, *Creativity, Annual Review of Psychology* (2009)
8. A.M. Isen, K.A. Daubman, G.P. Nowicki, Positive affect facilitates creative problem solving. J. Personality Soc. Psychol. **53**(6), 1122 (1987)
9. P. Jonathan, J.A. Russell, B.S. Peterson, The circumplex model of affect: an integrative approach to affective neuroscience, cognitive development, and psychopathology. Dev. Psychopathol. **17**(3), 715–734 (2005)
10. K.H. Kim, Can we trust creativity tests? A review of the torrance tests of creative thinking (TTCT). Creativity Res. J. **18**(1), 3–14 (2006)
11. N.W. Kohn, P.B. Paulus, Y. Choi, Building on the ideas of others: An examination of the idea combination process. J. Exp. Soc. Psychol. **47**(3), 554–561 (2011)
12. H. Lee, H.K. Kyung, Can speaking more languages enhance your creativity? Relationship between bilingualism and creative potential among Korean American students with multicultural link. Personality and Individ. Differ. **50**(8), 1186–1190 (2011)
13. C. Martindale, K. Anderson, K. Moore, A.N. West, Creativity, oversensitivity, and rate of habituation. Personality Individ. Differ. **20**(4), 423–427 (1996)
14. R.E. Mayer, 22 Fifty years of creativity research. Handb. Creativity **449** (1999)
15. K. Muldner, W. Burleson, Utilizing sensor data to model students' creativity in a digital environment. Comput. Hum. Behav. **42**, 127–137 (2015)
16. N. Murray, H. Sujan, E.R. Hirt, M. Sujan, The influence of mood on categorization: a cognitive flexibility interpretation. J. Pers. Soc. Psychol. **59**(3), 411–425 (1990)
17. J. Neter, W. Wasserman, M. Kutner, *Applied Linear Statistical Models: Regression, Analysis of Variance, and Experimental Designs*, (1990)
18. J.A. Plucker, Beware of simple conclusions: the case for content generality of creativity. Creativity Res. J. **11**(2), 179–182 (1998)
19. D.M. Rattner, *How to Use the Psychology of Space to Boost Your Creativity*, (2017)
20. N. Saeki, X. Fan, L. Van Dusen, A comparative study of creative thinking of American and Japanese college students. J. Creative Behav. **35**(1), 24–36 (2001)
21. P. Thagard, T.C. Stewart, The AHA! experience: creativity through emergent binding in neural networks. Cogn. Sci. **35**(1), 1–33 (2011)
22. E.P. Torrance, Torrance Tests of Creative Thinking: Norms-Technical Manual: Verbal Tests, forms a and b: Figural Tests, forms a and b. Personal Press, Incorporated, (1966)
23. E.P. Torrance, Torrance Tests of Creative Thinking: Verbal Tests, forms a and b, Figurals Tests, forms a and b: norms-technical manual. Personel Press/Ginn, Xerox Education Company, (1974)
24. E.P. Torrance, *The Search for Satori and Creativity* (Bearly Limited, Buffalo, NY, 1979)
25. E.P. Torrance, O. E. Ball, *Torrance Tests of Creative Thinking Streamlined (Revised) Manual, Figural A and B*. Scholastic Testing Service, (Bensenville, IL 1984)

Chapter 2
Using Body Signals and Facial Expressions to Study the Norms that Drive Creative Collaboration

J. Santuber, B. Owoyele, R. Mukherjee, S. K. Ghosh, and J. A. Edelman

Abstract Collaboration and creativity are consistently among the top-ranked values across societies, industries, and educational organizations. What makes collaboration possible is social norms. Group-based norms have played a key role in the evolution and maintenance of human ability to work and create together. We are not born collaborative-beings; it is the ability for social cognition and normativity that allows us to collaborate with others. Despite social norms ubiquity and pervasiveness—and being one of the most invoked concepts in social science—it remains unclear what are the underlying mechanisms to the extent to be one of the big unsolved problems in the field. To contribute to close this gap, the authors take an enactive-ecological approach, in which social norms are dynamic and context-dependent socio-material affordances for collaborative activity. Social norms offer the agent possibilities for collaborative action with others in the form of *pragmatic social cues*. The novelty of this research is the application of quantitative methods using computational models and computer vision to collect and analyze data on the *pragmatic social cues* of social norms in creative collaboration. Researchers will benefit from those methods by having fast and reliable data collection and analysis at a high level of granularity. In the present study, we analyzed the interpersonal synchrony of physiological signals and facial expressions between participants, together with the participant's perceived team cohesion. Despite the small size of the experiment, we could find correlations between signals and patterns that provide confidence in the feasibility of the methods employed. We conclude that the methods employed can be a powerful tool to collect and analyze data from larger groups and, therefore, shed some light on the—still not fully understood—underlying mechanisms of social normativity. The findings from the preliminary study are by no means conclusive, but serve as a proof of concept of the applicability of body signals and facial expressions to study social norms.

J. Santuber (✉) · B. Owoyele · J. A. Edelman
HPI-Stanford Design Thinking Research Program, Hasso Plattner Institute, Potsdam, Germany
e-mail: joaquin.santuber@hpi.de

R. Mukherjee · S. K. Ghosh
University of Potsdam, Potsdam, Germany

A. Przegalinska et al. (eds.), *Digital Transformation of Collaboration*,
Springer Proceedings in Complexity,
https://doi.org/10.1007/978-3-030-48993-9_2

17

Introduction

Collaboration is exclusively a human activity because of our capacity to create and regulate our actions by group-based norms. In contrast, other animals, even "our closest living primate relatives lack normative attitudes and therefore live in a non-normative socio-causal world structured by individual preferences, power relationships, and regularities" [1]. Humans are "normative animals". Even more, extensive literature shows how norms have played a key role in the evolution and maintenance of human cooperation, collaboration, social institutions, and culture [1–3]. Human normativity is the foundation of human's evolutionary game-changer: large-scale collaboration among unrelated strangers. This capacity for normativity lies at the core of "uniquely human forms of understanding and regulating socio-cultural group life" [1].

Social norms are ubiquitous and pervasive in human interactions. Despite that, they remain to be one of the main unsolved problems in social cognitive science. According to Fehr and Fischbacher,

> although no other concept is invoked more frequently in the social sciences, we still know little about how social norms are formed, the forces determining their content, and the cognitive and emotional requirements that enable a species to establish and enforce social norms [4].

Social norms are "standards of behavior that are based on widely shared beliefs how individual group members ought to behave in a given situation" [4]. Norms tell us "what to do" and "how to do" in different situations—regulate our actions when engaging with others in a team. Normative negotiations are often observed in team interactions in the form of questions such as "What are we doing?" and "How are we doing what we are doing?". Those questions seek an agreement on a constitutive norm (what is that we are doing) and regulative norms (how we do what we are doing). This study seeks to expand our understanding of team dynamics and collaboration by understanding its norms and the *social pragmatic cues* that afford collaboration.

However, the study of norms faces three challenges: (1) they are largely invisible, as they are implicit in the interaction of the participants, (2) norm-based environments are complex dynamic systems, they are not "simply being imposed on agents a priori" [5], and (3) norms are situated, embedded in a specific setting of a practice. Normativity is part of the "ongoing negotiation of identity and cultural meaning" of a community of practice [6].

In the first part of this paper, we lay out our theoretical approach that brings together (a) complex systems theory, (b) an enactive-ecological approach to cognition, and (c) social learning systems as a framework to studying social norms in the context of creative collaboration.

In the second part, we look at a preliminary study, which serves as a proof of concept to a quantitative approach to studying social norms by using body signals and facial expressions.

Theoretical Background

Norms as Media for Coupling of Systems: A Complex Systems Theory Approach

This study understands creativity as a process that emerges through collaboration. The creative process has its own existence in the form of a self-organizing and emergent system. The background theory is systems theory from biology, sociology, and creativity [7–9].

Creative collaboration: Norms as media for coupling of systems

Teams are self-organizing social systems that emerge from the effective coupling of psychic systems (team members). As a system of systems, teams have a precarious existence. This existence depends on the strength of the structural coupling of systems that is uncertain; it may happen or not. The coupling occurs through perception–action between *the team members and its environment.* To provide certainty to this coupling, *media*—an evolutionary artifact—needs to come in place to guide the perception and action of team members [7]. To our account, *norms are media that facilitates the structural coupling between team members that affords collaboration to emerge.* In other words, norms are cultural outcomes that increase our chances to survive collaboratively.

In social systems theory, language is the most common media that facilitates the coupling. When it comes to collaboration language alone is not enough. As shown above, social norms are the media that makes the structural coupling between team members possible. However, in everyday social interactions agents have to infer whether an act is normative from subtler, *social pragmatic cues* [1]. Doing that requires capacities for intersubjectivity or collective intentionality and shared values in a group. That is the "ability to share attention and mental states (e.g., intentions, goals) with conspecifics and thus to engage in shared intentional activities" [1]. Norms facilitate the emergence of a team's shared perception.

Norms in the Environment: An Enactive-Ecological Approach

Our perception and our actions are guided by norms present in our environment [10]. Norms are perceived by the agents through affordances [11]. Affordances, in a broad sense, offer the agent possibilities for action. In the case of norms, not only *what* actions but also *how* to act. Those affordances emerge from the interaction between the agent and the *socio-material environment.* It means the social interaction and cultural setting as well as the physical context in which collaboration emerges [12].

An enactive-ecological account of norms implies that norms are dynamic, emerging in the socio-material landscape. In this sense, "norms must also be understood as an embodied and situated practical sensitivity to the unfolding dynamics of the here-and-now contextual particularities of practices" [13]. Therefore, affordances are possibilities for skilled action depending on the competencies that the group of agents has. The different norms that emerge are defined by the practice and experiences of the team members and the setting in which they are situated—the rich socio-material landscape of affordances [14].

Norms in a Community of Practice

Norms belong to a particular practice, a form of life, a setting [10, 12, 15]. And here we understand teams, as social systems, to be a *social learning system* in the form of communities of practice [16]. According to Wenger, the engagement in a community of practice involves a dual meaning-making process through *participation* and *reification*. *Participation* is materialized through direct engagement in activities, conversations, and reflections; *reification* in the production of physical and conceptual artifacts—words, tools, concepts, stories that organize our participation [6].

Normativity is a form of reification; norms are conceptual artifacts that coordinate and anchor our perception and participation. They provide a common meaning to the shared experience as a team and community.

Social Norms as Solicitations to Collaborating

The authors take an enactive-ecological approach, in which social norms are dynamic and context-dependent socio-material affordances for collaborative activity. Social norms offer the agent possibilities for collaborative action with others in the form of *pragmatic social cues*. These social cues are context-dependent, and they belong to every practice.

Using Physiological Data and Facial Expressions Synchrony to Study Norms in Creative Collaboration: Individual and Team Level

The understanding of teams as complex systems implies a multidirectional causality of its dynamics. As in situated cognition, "one of the fundamental concepts is that cognitive processes are causally both social and neural. A person is obviously part of society, but causal effects in learning processes may be understood as bidirectional" [17]. The same applies to teams and individuals. The team behavior and its normative status affect the physiological responses of the team member. Likewise, the team member's physiological processes affect team behavior and team norms. The descriptive and normative accounts of reality are dependent phenomena and are causally related. That action–perception–reaction is part of the normative negotiation of that community.

In the paper "Socially Extended Cognition and Shared Intentionality", Lyre [18] offers a more detailed account of aspects of the environment that can provide the social cues for the coupling. He claimed that "virtually all mechanisms studied in social cognition [...] can be seen as potential coupling mechanisms of social extension" [18]. Building on his suggested list, we have included aspects from the enactive-ecological approach to the list, as well as a categorization of those mechanisms based on the normative (Table 2.1) and descriptive (Table 2.2) distinction. The normative mechanism for collaboration cannot be accessed directly. They need to be inferred from the descriptive *pragmatic social cues*. The descriptive mechanism for collaboration can be accessed directly. They are available in the socio-material environment in the form of *pragmatic social cues* that together constitute affordances to collaborate.

The *social pragmatic cues* of a specific practice can be inferred by integrating multiple descriptive, factual aspects of the socio-material environment. Specifically, intersubjective descriptive mechanisms need to be integrated into the situated descriptive to be able to *infer* the normative aspect of the social situation. The social norms that guide our perception and action are key to preserve the harmony—or chaos—of the community in which that collective experience is situated.

To capture this bidirectional causality—the individual and the team—this study leverages advanced approaches using digital technology for data collection and analysis. In this research in progress, we collected and analyzed physiological data using wearable devices, facial expressions using video recordings, and perceived team cohesion via self-assessment. In the study, participants engaged in a creative collaboration task provide a proof of concept for using body signals to understand the norms that drive creative collaboration.

Table 2.1 The normative mechanism for collaboration

Situated	Intersubjective
Cultural institutions	Shared goals
Social learning	Shared intentionality
Social norms	Co-operative action
Language	Communicative action
	Shared mental model

Table 2.2 Descriptive mechanisms for collaboration—*pragmatic social cues*

Situated	Intersubjective
Time	**Physiology**
Physical space	Gaze direction
Tools	Head pose
	Body posture
	Gestures
	Facial expression

To clearly frame the scope of this preliminary study, two research questions are addressed:

1. What are the right theoretical frameworks for the study of group-based norms in creative collaboration?
2. How can new computer-powered automatic data collection and analysis methods contribute to quantitatively study of social norms in creative collaboration?

Study Design

Participants

For this preliminary experiment 10 participants—6 females and 4 males—were recruited from a group of graduate students from a digital engineering institute in Germany. Regarding their profession, all of them are researchers in the field of computer science with a highly homogenous cultural background—northern European—and none of them was a native English speaker. The age of the participants was between 21 and 28 years and they had no previous experience working together. The participation was voluntary—not subject to any payment. The participants were paired based on their time availability, which resulted in four gender-diverse dyads and one dyad of females. All participants signed the corresponding informed consent form. All procedures performed in this study involving human participants were in accordance with the ethical standards of the institutional research committee and with the 1964 Helsinki declaration and its later amendments or comparable with ethical standards.

Fig. 2.1 Experimental setup

Procedure

The experiment was to collaboratively work with a wooden puzzle. The experiment was divided into three consecutive tasks, each one of them with the same 3-D wooden puzzle of a Dinosaur but with a different set of instructions (Fig. 2.1). The total length of the experiment was 45 min, including 5 min of baseline before the Tasks 1, 2, and 3. Every task lasted for 5 min and was followed by a 3 minutes break to fill a survey on perceived team cohesion (PTC) (see Table 2.3).

The following task instructions were given to the participants: For Task 1, the participants were given the puzzle and its cover (see Fig. 2.2), without any other

Table 2.3 Experiment procedure

Task 1—Situated normative status (5 min)	Task 2—discordant normative status (5 min)	Task 3—prescriptive normative status (5 min)
Participants were told to collaborate creatively and were provided with a 3-D wooden puzzle and its cover (see Fig. 2.1)	Participants received slightly different written instructions to assemble the puzzle, showed for 10 s Participant 1: The goal of this session is to assemble the puzzle through creative collaboration Participant 2: The goal of this session is to creatively explore different assembling of the pieces through collaboration	A visual guide on how to assemble the puzzle was provided (see Fig. 2.1)

Fig. 2.2 Dinosaur Set 1, cover sheet (left side) that served as visual instructions for task 1 and one of the pages of the assembly guide (right side) provided to the participants at the beginning of task 3

instruction. For Task 2, they were given two different written instructions to assemble the puzzle through creative collaboration. For Task 3, they were given no instructions but an assembly guide of the puzzle (see Fig. 2.2).

Data Collection

The data collected during the experiment consisted of perceived team cohesion, electrodermal activity (EDA), and heart rate (HR) and video, from which we can extract facial expressions of the participants. For the collection of EDA and HR data from the participants, we used the Empatica E4 wristband [19]. The data collection was done using a stationary setup; no audio-visual staff was present during the recording.

Perceived Team Cohesion

To measure the perception of the participants for each task, we use a self-report questionnaire. The "Perceived Team Cohesion Questionnaire" (PTCQ) has 10 questions and was answered individually by every participant after every task using a Likert scale (Table 2.4). The questionnaire was adapted from the paper "Physiological evidence of interpersonal dynamics in a cooperative production task" [20].

The changes in the perception of collaborative work collected with PTCQ serve two purposes. First, it was used as a validation measurement that the experiment design and intervention did actually generated a change—especially in Task 2—and that the change was perceived by the participants. The second purpose was to study correlations between PTC and synchrony of physiological signals and facial expressions.

Table 2.4 Perceived team cohesion questionnaire (PTCQ)

(1)	Based on what happened during the task right before…
(2)	I would like to interact with the other participant again
(3)	The other participant is a person I could see having as a friend
(4)	The other participant was warm
(5)	The interaction with the other participant went smoothly
(6)	I feel held back by the other participant
(7)	I do not fit in well working with this person
(8)	I felt uneasy with the other participant
(9)	How much did you want your group to perform well?
(10)	We did not have to rely on one another to complete the task

EDA

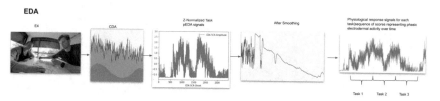

Fig. 2.3 Data processing of raw electrodermal activity data (EDA) included tonic extraction using continuous decomposition analysis (CDA) and data normalization. After that, the data corresponding to Tasks 1, 2, and 3 were manually extracted

Physiological Data

Electrodermal activity (EDA)

According to Boucsein [21], EDA refers to the electrical potential on the surface of the skin which is controlled by the sympathetic nervous system increases in sudomotor innervation, causing EDA to increase and perspiration to occur. Quick changes in EDA—arousals—are a response to stress, temperature, or exertion and have been frequently used in studies related to affective phenomena and stress [22]. In this preliminary study, synchrony between participants is calculated based on the similarity of arousal peaks, technically known as skin conductance response (SCR). The shape of an SCR—arousal—should typically last between 1 and 5 s, has a steep onset and an exponential decay, and reaches an amplitude of at least 0.01 μs [21].

For the collection of EDA data from the participants, we used the Empatica E4 wristband [19], which collects EDA at a frequency of 4 Hz by using two electrodes on the skin.

Before data analysis of SCR, EDA needs to be processed. For EDA, the raw data consists of phasic and tonic EDA. To study the synchronization between SCR—phasic—of two signals we need to extract the tonic. To extract it, the raw data for each participant was visualized using Ledalab and a continuous decomposition analysis was run [23]. Because of individual dependency of EDA and to avoid noise, we smoothed the data and normalized it using a z-score normalization on the signal (see Fig. 2.3).

In the phasic EDA sheet, we got two columns; one is the timestamp and the other is the amplitude of the signal. We cut the data manually in the interval of the desired time in seconds by the column of the timestamp. The data of two participants for each task were plotted in one graph to analyze the synchronization between both of them, as described in Section "Data Analysis".

Heart Rate (HR)

The second physiological measure is HR, which captures the difference between interbeat intervals (IBI) and is important in estimating vagal tone and parasympathetic nervous system activity.

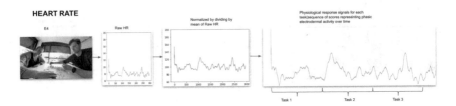

Fig. 2.4 Data processing of raw heart rate (HR) data normalization by mean extraction. After that, the data corresponding to Tasks 1, 2, and 3 were manually extracted

To collect HR, we used the Empatica E4 wristband that uses a photoplethysmography sensor to illuminate the skin and measures the light reflected by the presence of oxyhemoglobin. According to Garbarino and colleagues [19] with each cardiac cycle, the heart pumps blood to the periphery, changing the volume and pressure produced by the heartbeat which correlated to a change in the concentration of oxyhemoglobin.

Before the analysis of synchrony between participants, every HR raw data was normalized, dividing it by the mean, in order to reflect changes in HR relative to a baseline—the mean. We cut the data manually in the interval of the desired time in seconds corresponding to every task (Fig. 2.4). The data of two participants—for every dyad and for each task—were plotted in one graph to analyze the synchronization between both of them as described in Section "Data Analysis".

Facial Expressions

Our faces offer a rich source of *pragmatic social cues*. From facial expressions we communicate and infer emotions and intentions; they serve as a visual guide on how to act during social interactions and encounters with others [24]. Previous research on facial mimicry considers it a "basic facet of social interaction, theorized to influence emotional contagion, rapport, and perception and interpretation of others' emotional facial expressions" [25].

In this study, we use computer vision to analyze facial action units (FAU), a coding system based on the muscle of the face that is correlated with certain emotional states.

Video footage of the experiment was captured using a 360° video camera. A free capture of the face of each participant was extracted from the 360° video, for every task. Every video was analyzed using the open-source application Openface 2.0 [26] for automatic facial behavior analysis. Based on the software confidence output, noisy data was removed—less than 5 percent of total frames. The frequency of analysis is 30 frames per second, which provides a very rich and high-granularity data. The FAU analysis provides two values, presence and intensity for every FAU.

Before the synchrony analysis, the raw is coded into positive, negative, and neutral facial expression based on the values of FAU. Positive expression was coded if AU_12 is there, then it will indicate positive expression and the amplitude in those points will be the average of the intensity of AU_06 and AU_12. A negative expression was

FACIAL EXPRESSIONS

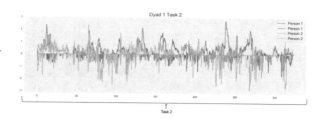

Fig. 2.5 Facial expressions were extracted using automatic facial behavior analysis using computer vision and coded into positive and negative emotions based on FAU presence and intensity. The intensity values were plotted together to analyze synchrony

coded if AU_15 and AU_01 are present and the amplitude in those points will be the average of the intensity of AU_15, AU_01, and AU_04. Since the second coding scheme is negative emotions, to make the value negative in the plot, we multiplied the set by −1. Positive, negative, and neutral data points for two participants for each task were plotted in one graph to analyze synchronization between both of them, as explained in Section "Data Analysis" (Fig. 2.5).

Data Analysis

Synchrony of Physiological Signals

Physiological interpersonal synchronization for every dyad was calculated using dynamic time warping (DTW) [27]. In particular, we analyzed the synchrony of EDA, HR signals and facial expressions—positive and negative separately—between two partners in a dyad. DTW is an algorithm for measuring the similarity between two temporal sequences that vary in time and speed. DTW provides the distance between the partners' physiological response signals for each task. Interpersonal influence in social interactions occurs normally within a five seconds timeframe [27]. For that reason, we enforced a locality constraint of five seconds while searching for the nearest points between the signals. As the frequency of the EDA signals is 4 Hz in our case, the constraint window consists of 20 samples within which we searched for the similarity. For HR, the five seconds windows corresponded to five data points. For the facial expressions, the window size corresponded to 150 data points—30 samples per second (Figs. 2.6 and 2.7).

Once we had the synchrony coefficient of every dyad for Tasks 1, 2, and 3, we looked for the correlation between the different signals and the perceived team cohesion (PTC) data from the survey using Pearson's correlation coefficient (PCC) [28]. The PCC range from 1 to −1 (positive to negative/inverse correlation).

The synchrony coefficient was normalized to z-scores and multiplied by −1, to ensure comparability with PTC.

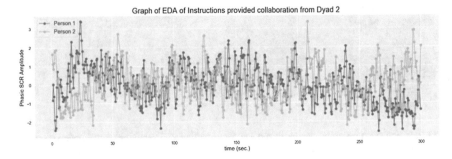

Fig. 2.6 EDA plot of signals from participants 1 and 2 corresponding to dyad 2 in task 3

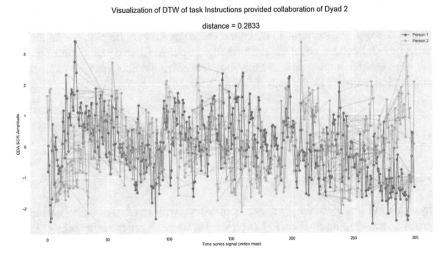

Fig. 2.7 DTW measures the distance between the two signals to provide a physiological synchronization coefficient based on EDA corresponding to dyad 2 in task 3

Perceived Team Cohesion Analysis

The data from the PTCQ were averaged between dyads and then normalized across tasks to z-scores.

Results

Because of the preliminary nature of this study, its main goal is to explore and test new methods to study social norms in collaboration. In the following paragraphs, we present some initial findings, making clear to the reader that the results are not meant to be conclusive but rather illustrative of the expected results (Fig. 2.8).

A general overview of the data shows consistency in the changes of interpersonal synchrony levels due to the experiment design and intervention. Of special interest to the authors was observing the changes experienced between Tasks 1 and 2 and between Tasks 2 and 3. Tasks 1 and 3 are designed to have a high level of normative agreement with a shared goal. In contrast, the intervention in Task 2 was meant to misalign the participant's goal by providing different instructions before the task. This served as a first validation of the intervention design and the methods employed.

Physiological synchrony and correlation between EDA sync and HR is positive and show a direction that could be further explored with a larger data set. EDA sync correlations values are not strong enough to be considered. Interestingly, EDA follows the tendency of negative correlation with negative facial expression synchrony. For HR sync, a positive correlation with PTC (r. 0.344) can be seen, which are also in agreement with the intervention design.

Facial expressions synchrony correlation, between positive and negative facial expression sync, are strongly negative. This correlation can be explained because of the experiment design of Tasks 1 and 3, agreement and shared goals versus Task 2, disagreement and misalignment, associated with more negative facial expression. In the same direction, the correlation between facial expressions and PTC is positive for positive facial expression and negative for negative expressions. The consistency between facial expression and PTC places facial expressions as a potential reliable indicator of normative agreement (Table 2.5).

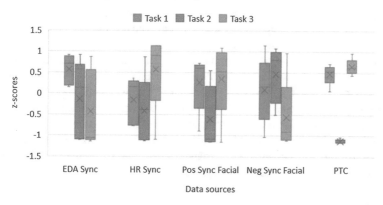

Fig. 2.8 Synchrony coefficients z-scores per index plus PTC

Table 2.5 Descriptive statistics and correlations

	Mean	Stdev	Min	Max	1	2	3	4	5
1. EDA Sync	10.281	7.634	1.637	32.044	1.00				
2. HR Sync	4.470	2.626	0.953	8.825	0.202	1.00			
3. Pos Face Sync	22.890	6.823	14.776	38.210	0.156	-0.032	1.00		
4. Neg Face Sync	38.938	11.907	18.456	59.975	-0.196	-0.259	-0.490	1.00	
5. PTC	4.183	0.723	2.300	4.850	0.004	0.344	0.439	-0.311	1.00

Discussion

As a preliminary study investigating the quantitative correlations between different levels of normative agreement, levels of perceived team performance, physiological synchronization, and facial expression synchrony the experiment provides feasibility proof for further research with a larger group of participants and more means of measure synchronization to understand social norms. Quantifying social norms opens up new opportunities for research. Norms are at the core of human collaboration, therefore, advancing our understanding of social norms is the key to any other form of collaboration.

The enactive-ecological as a theoretical framework provides a flexible yet rigorous and solid foundation from cognitive sciences and philosophy. A strong advantage of understanding group-based norms as socio-material affordances available in the form of *pragmatic social cues* broadens the scope of study of normativity in collaboration. It moves away from a "head-locked" approach to social norms and brings embodied and environmental aspects that can contribute to better understand the mechanism underlying our unique ability for normativity and collaboration. This approach opens up the scope of the socio-material environmental aspects that have not been traditionally considered in social studies.

In the same direction, new computer-powered data collection methods such as wearable devices have become unobtrusive and accessible. This technological advancement in sensing devices enables larger studies with cost-effective data collection. Regarding data analysis, open-source software and machine learning can provide fast and efficient methods for the analysis of large data sets. The preliminary experiment presented in this study would have taken great amounts of manpower in terms of time and knowledge for manual coding and labeling of data. Thanks to reliable and rigorous methods or methodologies, studies that integrate multiple data sources can contribute to mapping the bigger picture of group-based norms in collaboration.

Future Research and Limitations

Social norms are invisible and dynamics but people can navigate them quite successfully, thanks to pragmatic social cues found in the environment. This preliminary focused on the use of computer-powered automatic data collection and analysis methods as a means for the understanding of social norms. The study of social norms will remain incomplete if the research cannot integrate multimodalities of the *pragmatic social cues*. We believe quantitative methods can contribute to bridging that gap in a time and cost-effective manner. The limitations of such an extensive approach are the sacrifices of the phenomena of fidelity when compared to qualitative methods. Future research will focus on other social cognitive vehicles such as body posture, gestures, eye movement, head pose and communicative actions to gain a better understanding of social norms in a similar fashion people do. Adding these different pieces together can give us a map of the coupling structures of social norms.

References

1. M. Schmidt, H. Rakoczy, (eds.), *On The Uniqueness of Human Normative Attitudes*, ed. by K. Bayertz, N. Roughley. The Normative Animal? On the Anthropological Significance of Social, Moral and Linguistic Norms. (Oxford University Press, 2019)
2. R. Boyd, P.J. Richerson, Culture and the evolution of human cooperation. *Philos. Trans. R. Soc. Lond. Series B, Biol. Sci.* **364**(1533), 3281–3288 (2009). https://doi.org/10.1098/rstb.2009.0134
3. Michael Tomasello, Amrisha Vaish, Origins of human cooperation and morality. Annu. Rev. Psychol. **64**, 231–255 (2013). https://doi.org/10.1146/annurev-psych-113011-143812
4. E. Fehr, U. Fischbacher, Social norms and human cooperation. Trends Cogn. Sci. **8**(4), 85–90 (2004)
5. D. Hadfield-Menell, M.K. Andrus, G.K. Hadfield, Legible Normativity for AI Alignment: The Value of Silly Rules (2018). http://arxiv.org/pdf/1811.01267v1
6. E. Wenger, *Communities of Practice and Social Learning Systems: The Career of a Concept*, ed. by C. Blackmore. Social Learning Systems and Communities of Practice, vol. 14. (Springer London Ltd, England, 2010), pp. 179–198
7. T. Iba, An autopoietic systems theory for creativity. *Procedia Soc. Behav. Sci.* **2** (2010). https://doi.org/10.1016/j.sbspro.2010.04.071
8. Niklas Luhmann, The world society as a social system. Int. J. Gen Syst **8**(3), 131–138 (1982). https://doi.org/10.1080/03081078208547442
9. H.R. Maturana, F.J. Varela, Autopoiesis and Cognition, vol. 42. (Dordrecht, Springer Netherlands, 1980)
10. Erik Rietveld, Situated normativity: the normative aspect of embodied cognition in unreflective action. Mind **117**(468), 973–1001 (2008). https://doi.org/10.1093/mind/fzn050
11. J.J. Gibson, The *Ecological Approach to Visual Perception*. (Boston, MT, 1979)
12. J. Clancey, Situated Cognition: On Human Knowledge and Computer Representations. (Cambridge, Cambridge University Press (Learning in doing), 1997). http://www.loc.gov/cat dir/description/cam028/96035839.html
13. Presti Lo Patrizio, An ecological approach to normativity. Adapt. Behav. **24**(1), 3–17 (2016). https://doi.org/10.1177/1059712315622976

14. Ludger van Dijk, Erik Rietveld, Foregrounding sociomaterial practice in our understanding of affordances: the skilled intentionality framework. Front. Psychol. **7**, 1969 (2016). https://doi.org/10.3389/fpsyg.2016.01969
15. L. Wittgenstein, *Philosophical investigations. Philosophische Untersuchungen.* (Original 1953). (New York: Macmillan, 1958)
16. Etienne Wenger, *Communities of Practice* (Cambridge University Press, Cambridge, 1998)
17. Jeremy Roschelle, William J. Clancey, Learning as social and neural. Educ. Psychol. **27**(4), 435–453 (1992). https://doi.org/10.1207/s15326985ep2704_3
18. H. Lyre, Socially extended cognition and shared intentionality. Frontiers in Psych. **9**, 831 (2018)
19. M. Garbarino, M. Lai, D. Bender, R. Picard, S. Tognetti, Empatica E3–A wearable wireless multi-sensor device for real-time computerized biofeedback and data acquisition, in Proceedings of the International Conference on Wireless Mobile Communication and Healthcare (MobiHealth, 2014)
20. D. Mønster, D.D. Håkonsson, J.K. Eskildsen, S. Wallot, Physiological evidence of interpersonal dynamics in a cooperative production task. *Physiol. behav.r* **156**, 24–34 (2016). https://doi.org/10.1016/j.physbeh.2016.01.004
21. W. Boucsein, *Electrodermal Activity* (Springer Science and Business Media, LLC, 2012)
22. S. Taylor, N. Jaques, W. Chen, S. Fedor, A. Sano, R. Picard, Automatic identification of artifacts in electrodermal activity data, in Engineering in Medicine and Biology Conference, (2015)
23. M. Benedek, C. Kaernbach, A continuous measure of phasic electrodermal activity. J. Neurosci. Methods **190**, 80–91 (2010)
24. M. Martelli, J.M. Majib, D.G. Pelli, Are faces processed like words? A diagnostic test for recognition by parts. J. Vis. **5**, 58–70 (2005)
25. Daniel N. McIntosh, Spontaneous facial mimicry, liking and emotional contagion. Pol. Psychol. Bull. **37**(1), 31 (2006)
26. T. Baltrušaitis, A. Zadeh, Y.C. Lim, L. Morency, OpenFace 2.0: Facial Behavior Analysis Toolkit, in IEEE International Conference on Automatic Face and Gesture Recognition, 2018
27. P. Chikersal, M. Tomprou, Y. Kim, W.A. Williams L. Dabbish, Deep structures of collaboration: physiological correlates of collective intelligence and group satisfaction, in Proceedings of the ACM Conference on Computer Supported Cooperative Work and Social Computing (CSCW 2017), (2017)
28. J. Hernandez, I. Riobo, A. Rozga, G. Abowd, R. Picard, Using electrodermal activity to recognize ease of engagement in children during social interactions, in Proceedings of the International Joint Conference on Pervasive and Ubiquitous Computing (UbiComp 2014), (2014)

Chapter 3
Measuring Audience and Actor Emotions at a Theater Play Through Automatic Emotion Recognition from Face, Speech, and Body Sensors

Peter A. Gloor, Keith April Araño, and Emanuele Guerrazzi

Abstract We describe a preliminary experiment to track the emotions of actors and audience in a theater play through machine learning and AI. During a 40-min play in Zurich, eight actors were equipped with body-sensing smartwatches. At the same time, the emotions of the audience were tracked anonymously using facial emotion tracking. In parallel, also the emotions in the voices of the actors were assessed through automatic voice emotion tracking. This paper demonstrates a first fully automated and privacy-respecting system to measure both audience and actor satisfaction during a public performance.

Introduction

Emotion recognition has been widely studied for many years. Human emotion is a crucial element for communication and decision-making. The availability of emotion-rich data sources on many channels, along with recent advances in machine learning and deep learning have led to the development of various intelligent systems that are able to automatically recognize and interpret human emotions. In businesses for example, online retail systems are capable of analyzing emotional customer feedback to improve customer satisfaction [14]. In healthcare, the physical and emotional states of patients are monitored to automatically diagnose and prescribe the appropriate treatment [8]. Another application of emotion recognition is for safe driving through online monitoring of driver emotions [42].

P. A. Gloor (✉)
Massachusetts Institute of Technology, 77 Massachusetts Avenue, Cambridge, MA, USA
e-mail: pgloor@mit.edu

K. A. Araño
Politecnico di Milano, Via Lambruschini, 4/B, 20156 Milan, Italy

E. Guerrazzi
University of Pisa, Largo Lucio Lazzarino, 56122 Pisa, Italy

© Springer Nature Switzerland AG 2020
A. Przegalinska et al. (eds.), *Digital Transformation of Collaboration*,
Springer Proceedings in Complexity,
https://doi.org/10.1007/978-3-030-48993-9_3

Traditionally, emotion recognition research has been focused on analyzing unimodal data: speech signals [38], text data [43], facial expressions [18], and most recently, physiological signals [2]. However, emotions are complex cognitive processes with rich features that are difficult to infer with just a single modality [34]. Consequently, a number of studies have investigated the use of multimodal data and have shown that it can substantially improve the prediction of emotional states [41]. Moreover, the concept of cross-modal prediction in which shared representations are learned from multiple modalities to predict emotion from one modality to another has recently received growing interest from the research community.

One of the most widely known challenges in the field of emotion recognition is the difficulty of obtaining and labeling datasets to train prediction models. Cross-modal prediction addresses this problem by learning embeddings from one modality to predict another. For example, Albanie and colleagues [1] investigated the task of learning speech embeddings without access to any form of labeled audio data by exploiting a pre-trained face emotion recognition network, to reduce the dependence on labeled speech. Similarly, Li et al. [21] proposed a cross-modal prediction system between vision and touch that is capable of learning to see by touching, and learning to feel by seeing. Inter-modality dynamics, which models the interactions between different modalities and how they affect the expressed emotions of an individual, have also been investigated in earlier work [45]. The majority of these studies on using multimodal data, however, have been focused on recognizing emotions of a single individual, and little research has explored such inter-modality interactions between a group of individuals.

Motivated by these advances in multimodal emotion recognition, we investigate the correlations between the emotions depicted from facial expressions, speech, and physiological signals between two separate groups of individuals. In particular, we monitor the interaction between actors and audience in a theater performance. The contributions of this paper are as follows: We trained a face emotion recognition (FER) model and a speech emotion recognition (SER) model that are capable of predicting emotions from facial expressions and speech signals, respectively. We collected visual, audio, and physiological data from actors and audience of a theater performance which took place at the Landesmuseum in Zurich, Switzerland in spring of 2019. To the best of our knowledge, no such dataset has been collected before. We finally investigated the correlations between the emotions predicted by our deep learning models from the facial expressions, speech, and physiological signals that we have collected. Specifically, we analyzed the emotions of actors from their speech and physiological signals, and how these emotions translate to the facial expressions of the audience, investigating inter-modal and inter-personal dynamics.

Theoretical Background

Psychologists have proposed several theories categorizing different emotions that also account for age and cultural differences. One of the most widely applied emotion categorization frameworks is Paul Ekman's emotion model [12] where he classifies emotions into six basic categories: anger, happiness, fear, surprise, disgust, and sadness. Another universally recognized emotion classification system is the Circumplex model of affect [33], which is a two-dimensional model with valence describing the range of negative and positive emotions, and arousal depicting the active to passive scale of emotions. High valence and high arousal, for example, represent a pleasant feeling with high activation, which describes emotions such as happiness and excitement.

These emotions can be expressed in several ways: through facial expressions, speech, text, body language, or physiological signals. Among these modalities, facial expression is believed to be one of the most powerful and direct channels to convey human emotions in non-verbal communication [3, 35] while speech, on the other hand, is one of the most natural channels to transmit emotions in verbal interactions. These modalities differ in their potential in predicting emotional states as well as in their availability and usability under various circumstances [10]. Moreover, one modality can be influential in the recognition of another, which has been investigated in prior studies in cross-modal prediction [1, 21]. This can be useful in applications where one modality is utilized when the other is absent, such as in generating captions or labels for images [16] or in using vision to predict sounds [29].

Various methods have been proposed for recognizing emotions from faces, speech, and physiological signals. In face emotion recognition (FER), the current dominant techniques are deep neural networks (DNNs), such as convolutional neural networks (CNNs), which have been extensively used in diverse computer vision tasks that have resulted in several well-known CNN architectures such as AlexNet [19], VGG [36], VFF-face [30], and GoogleNet [39]. Similarly, in speech emotion recognition (SER), the recent breakthroughs in deep learning have led to the design of numerous DNN architectures such as variants of CNNs and long short-term memory (LSTM) networks, which have shown state-of-the-art performance in SER [20, 40]. In emotion recognition from physiological signals however, the majority of prior studies use classical algorithms such as support vector machines (SVM), random forests (RF), linear discriminant analysis (LDA), and K-nearest neighbors (KNN) [2]. Deep learning in this domain is still in its infancy, possibly due to the lack of large physiological emotion-labeled datasets necessary for training deep networks, contrary to FER where a substantial number of large datasets exist.

The aforementioned methods traditionally have not only dealt with unimodal data, but have also become popular in multimodal emotion recognition, in which the detection of emotion in each modality is a critical component for the success of the entire multimodal system. One of the key challenges in multimodal emotion recognition is to model the interactions between each modality (i.e. inter-modality dynamics) [25]. While novel approaches [45] have been proposed to address this problem, a majority of the earlier work has been focused on the inter-modality dynamics within a single individual. In psychology, numerous studies [31, 37] affirm that clearly another person's emotions do have an effect on our own actions, thoughts, and feelings. For instance, Ekman [11] highlights how one person's face may influence the emotional experience of another: "If B perceives A's facial expression of emotion, B's behavior toward A may change, and A's notice of this may influence or determine A's experience of emotion" [11]. In the field of multimodal emotion recognition, on the other hand, little research has been done to explore the inter-modality dynamics between individuals (i.e. inter-personal). This research aims to further understand such inter-modality and inter-personal effects. Through an empirical study, we investigate the correlations between emotions extracted from the speech and physiological signals of a group of individuals, and the emotions from the facial expressions of another group.

Methodology

Data Collection

We collected physiological, visual, and audio data from both actors and audience during a theater performance that took place in the Landesmuseum in Zurich, Switzerland on May 25, 2019. Through the Happimeter app [4] running on the smartwatch that the actors wore during the performance, we were able to gather the activation, pleasure, and stress levels of the actors. The Happimeter runs a trained machine learning model that is capable of predicting such emotions from the physiological signals that are collected from the sensors of the smartwatch. Through a video camera that was set-up inside the theater, we captured the faces of the audience and the voices of the actors during the entire performance which lasted for about 40 min.

Considering the number of smartwatches available as well as the privacy issues imposed by the collection of sensor data, we opted to collect the physiological signals from the smaller group of individuals—the actors, which consisted of eight individuals. Moreover, as a theater etiquette, loud whispers and conversations in the audience are discouraged, hence, speech data was only collected from the actors. Facial expressions, on the other hand, were recorded from the audience, which consisted of about 40 people. Table 3.1 summarizes the data collected during the theater performance.

Table 3.1 Summary of the collected data from the theater performance

Modality	Emotions	Group
Facial expressions	Anger, fear, happiness, sadness	Audience
Physiological signals	Activation, pleasure, stress	Actors
Speech signals	Anger, fear, happiness, sadness	Actors

Table 3.2 FER training set

Emotions	Dataset				Total
	CK+	JAFFE	BU-3DFE	FacesDB	
Happy	69	31	77	36	213
Sad	28	31	88	36	183
Angry	45	30	94	35	204
Fearful	25	32	92	36	185
Total	167	124	351	143	**785**

Model Implementation

Face Emotion Recognition

Our FER model, which has a prediction accuracy of 74.9%, has been trained on a combination of multiple datasets: CK+ [23], JAFFE [24], BU-3DFE [44], and FacesDB [27]. The cardinality of each emotion type in these datasets is summarized in Table 3.2. Our model uses a VGG16 [36] CNN architecture that was pre-trained on ImageNet. We freeze the layers except for the last four layers of this pre-trained model. We use SGD as an optimizer with a learning rate of 0.01 and a Softmax activation function in the dense output layer of the network. All of the detected faces from the camera were resized to 100×100 as input to the VGG16 model. Since VGG16 expects three input channels, we extend the images into three dimensions by using the same values for red, green, and blue (i.e. grayscale).

Using the face_recognition python package which is based on the dlib machine learning library [17], we detect the faces from the captured images on the camera. We then label all the recognized faces with the emotions happy, angry, sad, and fearful, using our trained FER model. The probabilities of the emotion classes are obtained from the Softmax layer of our FER network. Figure 3.1 illustrates a snapshot of the video captured by the camera, showing the emotion-labeled faces as predicted by our FER model.

Speech Emotion Recognition

Our SER model, which has a prediction accuracy of 71.01%, has been trained on a combination of multiple datasets containing 3–5 s of emotion-labeled audio files: RAVDESS [22], SAVEE [15], CREMA-D [7], IEMOCAP [6], TESS [9], and EMODB [5]. The cardinality of each emotion type in these datasets is summarized in Table 3.3. Using python's Librosa [26] library, we extract the Mel-frequency cepstral coefficients (MFCCs) from each audio file with a sampling rate of 44,100 Hz, a fast Fourier transform (FFT) window of 2048, and hop length of 512 samples. This

Fig. 3.1 Emotion-labeled faces detected by our FER model (face blurred for privacy reasons)

Table 3.3 SER training set

Emotions	Dataset						Total
	RAVDESS	SAVEE	CREMA-D	IEMOCAP	TESS	EMODB	
Happy	376	60	1271	595	400	72	2774
Sad	376	60	1271	1084	400	62	3253
Angry	376	60	1271	1103	399	128	3337
Fearful	376	60	1271	40	399	68	2214
Total	1504	240	5084	2822	1598	330	**11,578**

implementation uses the Hann window function on the signal frames and performs a short-time Fourier transform (STFT) to calculate the frequency spectrum. We extract a total of 40 MFCCs, excluding the zeroth coefficient as it represents the average log-energy of the signal, which carries limited speech information [28]. We then feed this MFCC feature vector into our LSTM: a five-layer network with one input layer, three hidden layers, and one dense output layer with a Softmax activation function.

Using the video captured by the camera, we extract the corresponding audio data by converting the mp4 into a wav file format. The entire audio stream was split, with a chunk length of 4 s, since our SER prediction model was trained on audio data with a similar average time duration. We then label each of the 4-s audio with the emotions happy, angry, sad, and fearful, using our trained SER model. The probabilities of the emotion classes are obtained from the Softmax layer of our SER network.

Physiological Emotion Recognition

We use the machine learning model that is deployed in the Happimeter [4] app to label the emotions from the physiological signals, that is, a physiological emotion recognition (PER). Signals were collected by the sensors of the smartwatch. The model processes physiological (e.g. movement, heart rate, etc.) and environmental (noise, weather, etc.) variables as inputs to a classifier. It uses Scikit-learn's [32] gradient boosting algorithm with a learning rate of 0.1 and a maximum depth of eight nodes in each tree. This machine learning model, which currently has a prediction accuracy of 79%, has been trained with the data that has been acquired from the users of the app in the last three years. Using this trained model, the data collected from the smartwatches that were worn by the actors were labeled with values ranging from 0 to 2 to indicate the levels of activation, pleasance, and stress.

Correlation Analysis

We compare the predicted emotions from the voices and physiological signals of the actors to the emotions from the facial expressions of the people in the audience. We merge the predictions from our SER (actors) and FER (audience) models based on the closest timestamp and perform a rolling window calculation (i.e. simple moving average) using different time windows to filter out noise and expose the under-lying properties of the curves. Subsequently, we perform a correlation analysis using Pearson's correlation coefficient (see Equation below), where n is the sample size, x_i and y_i are the individual sample points i, and $\bar{x}$ and $\bar{y}$ are the sample mean. The same process is followed to compare the emotions from the PER (actors) and the FER (audience) model. We also analyze the physiological signals (i.e. heartrate and movement) from the actors and examine their correlations with the emotions portrayed from the faces of the audience.

$$r_{xy} = \frac{\sum_{i=1}^{n}(x_i - x)(y_i - y)}{\sqrt{\sum_{i=1}^{n}(x_i - \bar{x})^2}\sqrt{\sum_{i=1}^{n}(y_i - \bar{y})^2}}$$

Results

PER Versus FER

Figure 3.2 shows the levels of activation, pleasance, and stress of the actors (as measured by the Happimeter app) and the four emotions of the audience (as measured

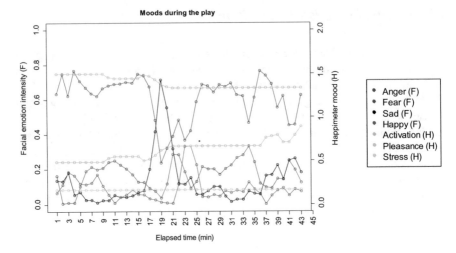

Fig. 3.2 Emotions from the Happimeter and the FER model

by the FER model) throughout the entire theater performance. As we can see, the pleasance of the actors went down as the play progressed, while their activation went up. The correlation values and the level of significance between these emotions are illustrated in Fig. 3.3 (* <0.05, ** <0.01, *** <0.001).

Fig. 3.3 Correlations between the emotions from the Happimeter (actors, about 900 measurements) and the FER model (audience, about 600 measurements)

	angry	fear	sad	happy	activation	pleasance
angry	1	-0.11	*** -0.74	-0.18	* -0.31	** 0.39
fear		1	* -0.37	-0.15	0.18	* -0.32
sad			1	-0.27	0.23	-0.2
happy				1	-0.1	0.06
activation					1	*** -0.9
pleasance						1

We find that activation of the actors and anger of the audience is negatively correlated ($r = -0.31*$). This means that the more excited the actors are, the less angry the audience is. We do not really assume that the audience is "angry", rather their facial expressions showed something that our FER interpreted as "angry". As we only had these four emotions labeled in this initial analysis, other emotions such as "surprise" or "insight" might be subsumed into the "angry" emotion, as the FER system might assign these emotions also the "angry" label. Similarly, we find that the higher the pleasance of the actors is, the less "fearful" the audience is ($r = -0.32*$). Somewhat counterintuitively we also find that the higher the pleasance of the actors is, the more angry the audience ($r = 0.39**$) is. This combination of correlations indeed suggests that the "surprise" facial expression might be similar to the "anger" facial expression and has been recognized as such by the FER.

Sensor Data Versus FER

In order to investigate the possible correlations between raw sensor data (as captured by the smartwatch) and the FER model, we collected and analyzed the data from the smartwatches worn by the actors. Figure 3.4 shows the average levels of movement (computed as the sum of the absolute values of accelerometers value along x, y, and z-axis), heartrate (beats per minute, BPM), and noise level (as measured by microphone). The correlation values and the level of significance between these emotions are illustrated in Fig. 3.5.

As Fig. 3.5 shows, the facial expression recognized as "angry" is negatively correlated to the average movement, that is, the less the actors move, the more "angry" the audience gets.

Sensor parameters (avg) and audience mood during the play

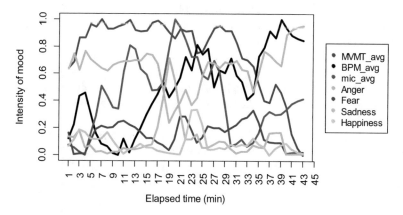

Fig. 3.4 Sensor data (average) comparison with the FER model

Fig. 3.5 Correlations between sensor data (average, about 2150 measurements) and the FER model

	MVMT_avg	BPM_avg	mic_avg	angry	fear	sad	happy
MVMT_avg	1	*** -0.52	0.2	** -0.42	* 0.36	0.13	0.18
BPM_avg		1	-0.06	0.13	-0.26	0.06	-0.05
mic_avg			1	-0.28	-0.03	* 0.3	-0.01
angry				1	* -0.34	*** -0.71	* -0.32
fear					1	-0.2	-0.21
sad						1	-0.07
happy							1

Similarly, we find that the higher the standard deviation in movement of the actors, the "angrier" expressions ($r = 0.32^*$) and the less "fear" expressions ($r = -0.34^*$) are recognized by the FER. This means that differences in movement among the actors trigger emotional reactions by the audience.

FER Versus SER

Figure 3.6 shows the plots of the probabilities of the emotion "anger" as measured by our FER and SER models using a rolling time window of 30 s, 1 min, and 5 min. As foreseen, a smoother curve is achieved with a longer time window. In Fig. 3.7 the plots of the probabilities of all four emotions between the actors (as predicted by the SER) and the audience (as predicted by the FER) with a rolling window of one minute are displayed. The corresponding correlation matrix showing the correlation values and the level of significance is displayed in Fig. 3.8. Only significant correlations between the emotions of the audience and actors (i.e. FER vs. SER predictions) are highlighted.

As the correlation matrix in Fig. 3.8 shows, "fear" in the faces of the audience is positively correlated with "anger" in the voice of the actors. "Anger" in the faces of the audience is positively correlated with "happiness" in the voice of the actor, which again suggested that "surprise" of the audience is also subsumed in this emotion.

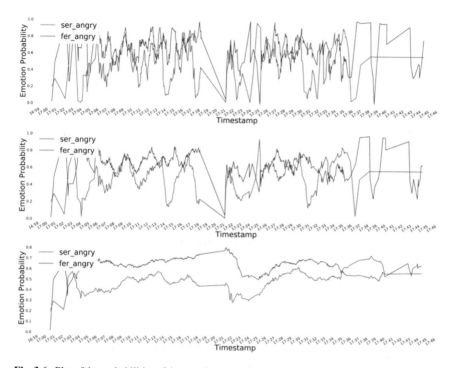

Fig. 3.6 Plot of the probabilities of the emotion anger from the FER and SER models using different time windows

Discussion

Emotions from Faces of the Audience Versus Voices of Actors

By taking into consideration a balance between filtering random noise or variations and preserving the original data, we chose a time window of 1 min to smoothen the time series predictions as can be observed from the plots in Fig. 3.6. Using this chosen time window, we see some obvious correlations between the emotions from the audience and the actors (see Fig. 3.8). A graphical summary of the correlations is shown in Fig. 3.9, which is based on the correlation matrix in Fig. 3.8.

For the emotion "anger", there is a statistically significant negative correlation between the audience and the actors. Interestingly, there is a statistically significant positive correlation between the "happiness" from the actors and "anger" from the audience. This implies that when there is "anger" from the actors, the audience feels less of the same emotion and similarly, when there is "happiness" from the actors, there is a higher intensity of "anger" from the audience.

The "anger" expressed by the voices of the actors is positively correlated with "fear" from the audience, which appear to be logical and can possibly infer that the

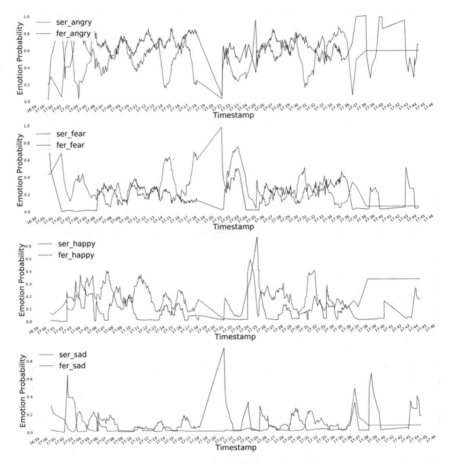

Fig. 3.7 Plot of the probabilities of the four emotions from the FER and SER models using a rolling window of 1 min

actors can effectively elicit "fear" from the audience by demonstrating "anger" in their voices. Consistent with such behavior, there is also a statistically significant negative correlation between the "happiness" from the actors and "fear" from the audience, implying that the audience feels less "fear" when the actors exhibit "happiness".

A statistically significant positive correlation is also present between "fear" from the actors and "sadness" from the audience. This may suggest that members of the audience are sympathetic, and they empathize with the "fear" from the actors by feeling "sad". Consistent with such observation, there is also a statistically significant negative correlation between "happiness" from the actors and "sadness" from the audience, which suggests that the actors effectively managed to make the audience feel less "sad" by showing "happiness" through their voices.

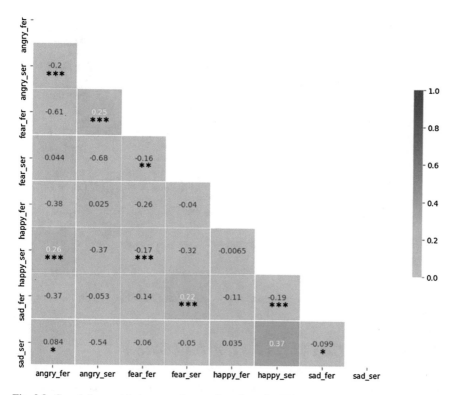

Fig. 3.8 Correlation matrix between the emotions from the FER ($N = 592$) and SER ($N = 684$) models

Emotions from Happimeter Versus Faces of the Audience

Based on the same considerations as discussed in section "Emotions from Faces of the Audience Versus Voices of Actors", we chose a time window of 3 min to smoothen the time series predictions of Happimeter and FER as shown in the plots in Fig. 3.6. In this plot, the actors' emotions "pleasance" and "activation" are compared with the audience's "angry", "fear", "sad", and "happy" facial expressions.

We find that the variable "angry" is negatively correlated to the "activation" of Happimeter and positively correlated to "pleasance". This seems to suggest that the angry emotion is covering another emotion (maybe "surprise") as it leads the audience to be more agitated. As expected, the audience variable "fear" is negatively correlated to the actors' "pleasance".

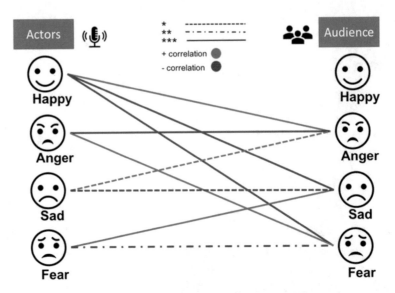

Fig. 3.9 Significant correlations between the actors and the audience based on the FER and SER correlation matrix

Actors' Sensor Data Versus Faces of the Audience

We find that an increase of the average "movement" of the actors leads to a decrease of "angry" emotions among the audience, in accordance with the discussion in section "Emotions from Happimeter Versus Faces of the Audience", but also to an increase of the "fear" emotion. Moreover, an increase of the average sound level measured with the microphone is positively correlated to the "sadness" of the audience. We assume that this is directly related to the theater piece which was played in this analysis, where tragic experiences of the protagonist are presented.

We also observed that an increase in the variance of movements leads to an increase in the anger of the audience, while decreasing their fear. This might be related to one actress walking among the audience, triggering some anger and fear of spectators of being called out.

A graphical summary of the correlations discussed in sections "Emotions from Happimeter Versus Faces of the Audience" and "Actors' Sensor Data Versus Faces of the Audience" is shown in Fig. 3.10, which is based on the correlation matrices given in Figs. 3.5 and 3.8.

Conclusions and Future Work

One of the main restrictions of the analysis described in this paper is that the FER we used is only capable of recognizing the four emotions: happy, sad, fear, and anger, potentially leading to over-recognition of fear and anger. In the revised version of the

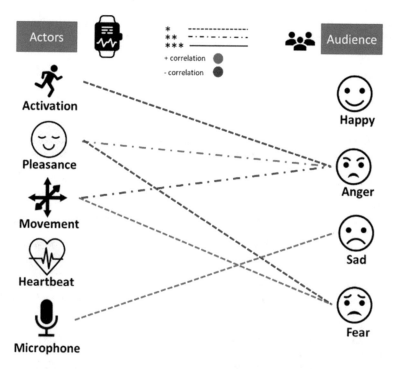

Fig. 3.10 Significant correlations between the actors and the audience based on the PER versus FER and sensor data versus FER correlation matrices

FER which has been developed in the meantime, we have included the two additional emotions of the Ekman model, surprise and disgust, which in more recent work have shown increased recognition accuracy and emotion coverage.

Nevertheless, we are convinced that the system described in this paper has illustrated the potential of our approach of automatically measuring audience and artist emotions at public events. We are currently extending our system for using it at other artistic events such as concerts and other public events. In particular, this includes giving immediate feedback to participants about their emotions, and combining sound input from other sources such as smartphones with the Happimeter and the video input from the webcam. Our ultimate goal will be to identify the emotions that will lead to optimal experiences for both performers and the audience. Mirroring back this behavior [13] to performers will allow them to better understand the impact their own emotions have on their audience, and thus to improve their artistic performance and skills.

Acknowledgements The authors thank Samuel Schwarz and Garrick Lauterbach from Digital-buehne.ch for providing the venue and supporting them during the performance. They are also grateful to Jannik Roessler for his invaluable support during data collection.

All procedures performed in studies involving human participants were in accordance with the ethical standards of the institutional/ and or national research committee with the Helsinki declaration and its later amendments or comparable with ethical standards. Consent by the organizers and participants in the Zurich experiment to be recorded has been given.

References

1. S. Albanie, A. Nagrani, A. Vedaldi, A. Zisserman, Emotion recognition in speech using cross-modal transfer in the wild, in *MM 2018—Proceedings of the 2018 ACM Multimedia Conference* (2018), pp. 292–301, https://doi.org/10.1145/3240508.3240578
2. M. Ali, A.H. Mosa, F.Al Machot, K. Kyamakya, A review of emotion recognition using physiological signals. Ann. Telecommun. **109**(3–4), 303–318 (2018). https://doi.org/10.1007/978-3-319-58996-1
3. N. Ambady, M. Weisbuch, Nonverbal behavior, in *Handbook of Social Psychology*, vol. 1, 5th ed. (Wiley, Hoboken, NJ, US, 2010), pp. 464–497
4. P. Budner, J. Eirich, P.A. Gloor, "Making You Happy Makes Me Happy"—Measuring Individual Mood with Smartwatches (Aristotle 2004) (2017), pp. 1–14, http://arxiv.org/abs/1711.06134
5. F. Burkhardt, A. Paeschke, M. Rolfes, W. Sendlmeier, B. Weiss, A database of German emotional speech, in *9th European Conference on Speech Communication and Technology*, vol. 5 (2005)
6. C. Busso, M. Bulut, C.-C. Lee, A. Kazemzadeh, E. Mower, S. Kim, S.S. Narayanan, IEMOCAP: interactive emotional dyadic motion capture database. Lang. Resour. Eval. **42**(4), 335 (2008). https://doi.org/10.1007/s10579-008-9076-6
7. H. Cao, D.G. Cooper, M.K. Keutmann, R.C. Gur, A. Nenkova, R. Verma, CREMA-D: crowd-sourced emotional multimodal actors dataset. IEEE Trans. Affect. Comput. **5**(4), 377–390 (2014). https://doi.org/10.1109/TAFFC.2014.2336244
8. M. Chen, Y. Zhang, M. Qiu, N. Guizani, Y. Hao, SPHA: smart personal health advisor based on deep analytics. IEEE Commun. Mag. **56**(3), 164–169 (2018). https://doi.org/10.1109/MCOM.2018.1700274
9. K. Dupuis, M. Pichora-Fuller, Recognition of emotional speech for younger and older talkers: behavioural findings from the toronto emotional speech set. Can. Acoust. Acoust. Can. **39**, 182–183 (2011)
10. M. Egger, M. Ley, S. Hanke, Emotion recognition from physiological signal analysis: a review. Electron. Notes Theor. Comput. Sci. **343**, 35–55 (2019). https://doi.org/10.1016/j.entcs.2019.04.009
11. P. Ekman, W.V. Freisen, S. Ancoli, Facial signs of emotional experience. J. Pers. Soc. Psychol. **39**(6), 1125–1134 (1980). https://doi.org/10.1037/h0077722
12. P. Ekman, W.V. Friesen, Constants across cultures in the face and emotion. J. Personal. Soc. Psychol. US: Am. Psychol. Assoc. (1971). https://doi.org/10.1037/h0030377
13. P. Gloor, A.F. Colladon, G. Giacomelli, T. Saran, F. Grippa, The impact of virtual mirroring on customer satisfaction. J. Bus. Res. **75**, 67–76 (2017)
14. W. Hong, C. Zheng, L. Wu, X. Pu, Analyzing the relationship between consumer satisfaction and fresh e-commerce logistics service using text mining techniques. Sustainability (Switzerland) **11**(13), 1–16 (2019). https://doi.org/10.3390/su11133570
15. P. Jackson, S. Ul haq, *Surrey Audio-Visual Expressed Emotion (SAVEE) Database* (2011)
16. A. Karpathy, F.-F. Li, Deep Visual-Semantic Alignments for Generating Image Descriptions. *CoRR, abs/1412.2* (2014), http://arxiv.org/abs/1412.2306
17. D.E. King, Dlib-ml: a machine learning toolkit. J. Mach. Learn. Res. **10**, 1755–1758 (2009)
18. B.C. Ko, A brief review of facial emotion recognition based on visual information. Sensors (Switzerland) **18**(2) (2018). https://doi.org/10.3390/s18020401

19. A. Krizhevsky, I. Sutskever, G.E. Hinton, ImageNet classification with deep convolutional neural networks, in *Proceedings of the 25th International Conference on Neural Information Processing Systems,* vol. 1 (Curran Associates Inc., USA, 2012), pp. 1097–1105, http://dl.acm.org/citation.cfm?id=2999134.2999257

20. J. Lee, I. Tashev, High-level feature representation using recurrent neural network for speech emotion recognition, in *Proceedings of the Annual Conference of the International Speech Communication Association, INTERSPEECH, 2015-Janua* (2015), pp. 1537–1540

21. Y. Li, J.-Y. Zhu, R. Tedrake, A. Torralba, *Connecting Touch and Vision via Cross-Modal Prediction, (d)* (2019), http://arxiv.org/abs/1906.06322

22. S.R. Livingstone, F.A. Russo, The Ryerson audio-visual database of emotional speech and song (RAVDESS) (2018), https://doi.org/10.5281/zenodo.1188976

23. P. Lucey, J. Cohn, T. Kanade, J. Saragih, Z. Ambadar, I. Matthews, The Extended Cohn-Kanade Dataset (CK+): a complete dataset for action unit and emotion-specified expression, in *2010 IEEE Computer Society Conference on Computer Vision and Pattern Recognition—Workshops, CVPRW 2010* (2010), https://doi.org/10.1109/CVPRW.2010.5543262

24. M. Lyons, M. Kamachi, J. Gyoba *The Japanese Female Facial Expression (JAFFE) Database* (Zenodo, 1998), https://doi.org/10.5281/zenodo.3451524

25. C. Marechal, D. Mikołajewski, K. Tyburek, P. Prokopowicz, L. Bougueroua, C. Ancourt, K. Węgrzyn-Wolska, Survey on AI-based multimodal methods for emotion detection, in *Lecture Notes in Computer Science (Including Subseries Lecture Notes in Artificial Intelligence and Lecture Notes in Bioinformatics),* vol. 11400 (2019), pp. 307–324, https://doi.org/10.1007/978-3-030-16272-6_11

26. B. McFee, C. Raffel, D. Liang, D. Ellis, M. McVicar, E. Battenberg, O. Nieto, librosa: audio and music signal analysis in python, in *Proceedings of the 14th Python in Science Conference,* (Scipy) (2015), pp. 18–24, https://doi.org/10.25080/majora-7b98e3ed-003

27. J. Mena-Chalco, R. Marcondes, L. Velho, *Banco de Dados de Faces 3D: IMPA-FACE3D* (2008)

28. D. Nandi, K. Rao, Language identification using excitation source features (2015), https://doi.org/10.1007/978-3-319-17725-0

29. A. Owens, P. Isola, J.H. McDermott, A. Torralba, E.H. Adelson, W.T. Freeman, Visually Indicated Sounds. *CoRR, abs/1512.0* (2015), http://arxiv.org/abs/1512.08512

30. O.M. Parkhi, A. Vedaldi, A. Zisserman, *Deep Face Recognition* (Section 3) (2015), pp. 41.1–41.12, https://doi.org/10.5244/c.29.41

31. B. Parkinson, How emotions affect other people. Emot. Res. (2014)

32. F. Pedregosa, G. Varoquaux, A. Gramfort, V. Michel, B. Thirion, O. Grisel, É. Duchesnay, Scikit-learn: machine learning in python. J. Mach. Learn. Res. **12**, 2825–2830 (2011). http://dl.acm.org/citation.cfm?id=1953048.2078195

33. J. Posner, J. Russell, B. Peterson, The circumplex model of affect: An integrative approach to affective neuroscience, cognitive development, and psychopathology. Dev. Psychopathol. **17**, 715–734 (2005). https://doi.org/10.1017/S0954579405050340

34. J.L. Qiu, W. Liu, B.L. Lu, Multi-view emotion recognition using deep canonical correlation analysis, in *Lecture Notes in Computer Science (Including Subseries Lecture Notes in Artificial Intelligence and Lecture Notes in Bioinformatics), 11305 LNCS* (2018), pp. 221–231, https://doi.org/10.1007/978-3-030-04221-9_20

35. N. Rule, N. Ambady, First impressions of the face: predicting success. Soc. Pers. Psychol. Compass **4**(8), 506–516 (2010). https://doi.org/10.1111/j.1751-9004.2010.00282.x

36. K. Simonyan, A. Zisserman, *Very Deep Convolutional Networks for Large-Scale Image Recognition* (2014)

37. R. Smith, A. Alkozei, W. Killgore, How do emotions work? Front. Young Minds **5** (2017). https://doi.org/10.3389/frym.2017.00069

38. M. Swain, A. Routray, P. Kabisatpathy, Databases, features and classifiers for speech emotion recognition: a review. Int. J. Speech Technol. **21**(1), 93–120 (2018). https://doi.org/10.1007/s10772-018-9491-z

39. C. Szegedy, W. Liu, Y. Jia, P. Sermanet, S.E. Reed, D. Anguelov, A. Rabinovich, Going Deeper with Convolutions. *CoRR, abs/1409.4* (2014), http://arxiv.org/abs/1409.4842

40. G. Trigeorgis, F. Ringeval, R. Brueckner, E. Marchi, M.A. Nicolaou, B. Schuller, S. Zafeiriou, Adieu features? End-to-end speech emotion recognition using a deep convolutional recurrent network, in *ICASSP, IEEE International Conference on Acoustics, Speech and Signal Processing—Proceedings, 2016-May* (2016), pp. 5200–5204, https://doi.org/10.1109/ICASSP. 2016.7472669

41. M.A. Ullah, M.M. Islam, N.B. Azman, Z.M. Zaki, An overview of multimodal sentiment analysis research: opportunities and difficulties, in *2017 IEEE International Conference on Imaging, Vision and Pattern Recognition, IcIVPR 2017* (2017), https://doi.org/10.1109/ICI VPR.2017.7890858

42. E. Vasey, S. Ko, M. Jeon, In-vehicle affect detection system: identification of emotional arousal by monitoring the driver and driving style, in *Adjunct Proceedings of the 10th International Conference on Automotive User Interfaces and Interactive Vehicular Applications* (ACM, New York, NY, USA, 2018), pp. 243–247, https://doi.org/10.1145/3239092.3267417

43. A. Yadollahi, A.G. Shahraki, O.R. Zaiane, Current state of text sentiment analysis from opinion to emotion mining. ACM Comput. Surv. **50**(2), 1–33 (2017). https://doi.org/10.1145/3057270

44. L. Yin, X. Wei, Y. Sun, J. Wang, M.J. Rosato, A 3D facial expression database for facial behavior research, in *Proceedings of the 7th International Conference on Automatic Face and Gesture Recognition* (IEEE Computer Society, Washington, DC, USA, 2006), pp. 211–216, http://dl.acm.org/citation.cfm?id=1126250.1126340

45. A. Zadeh, M. Chen, S. Poria, E. Cambria, L.-P. Morency, Tensor fusion network for multimodal sentiment analysis (2018), pp. 1103–1114, https://doi.org/10.18653/v1/d17-1115

Chapter 4
Measuring Moral Values
with Smartwatch-Based Body Sensors

Lirong Sun and Peter A. Gloor

Abstract In this research project we predict the moral values of individuals through their body movements measured with the sensors of a smartwatch. The personal moral values are assessed using the Schwartz value theory, which proposes two dimensions of universal values (open to change versus conservative, self-enhancement versus self-transcendence). Data for all variables are gathered through the Happimeter, a smartwatch-based body-sensing system. Through multilevel mixed-effects generalized linear models, our results show that sensor and mood factors predict a person's values. We utilized three methods to investigate the relationship between the Big Five personality traits (OCEAN: openness, conscientiousness, extraversion, agreeableness, and neuroticism) of a person and their Schwartz values. This research highlights the use of recent technological advances for studying a person's values from an integrated perspective, combining body sensors and mood states to investigate individual behaviour and team cooperation.

Introduction

Human behaviour is driven by conflicting emotions. To better understand the interaction of different human emotions, researchers have started using sensors for automatic recognition of individual traits, including happiness, physical/psychological health, satisfaction, and so forth [16]. Yet little research so far has addressed how to predict values through the lens of sensing technology. We know that values are

All procedures performed in studies involving human participants were in accordance with the ethical standards of the institutional and/or national research committee with the Helsinki declaration and its later amendments or comparable with ethical standards.

L. Sun · P. A. Gloor (✉)
MIT Center for Collective Intelligence, 245 First Street, Cambridge, MA 02142, USA
e-mail: pgloor@mit.edu

L. Sun
University of Chinese Academy of Sciences, Haidian District, Beijing 100190, China

© Springer Nature Switzerland AG 2020
A. Przegalinska et al. (eds.), *Digital Transformation of Collaboration*,
Springer Proceedings in Complexity,
https://doi.org/10.1007/978-3-030-48993-9_4

linked to behaviours, encouraging individuals to act in accordance with their values [25], and body sensors are the most honest way to pick up behaviours. In this regard, this demonstrates the feasibility of predicting values of a person with data that are collected by sensors.

In this study, we aim to advance an integrative view to study a person's values in terms of openness to change versus conservation, and self-enhancement versus self-transcendence, based on the Schwartz theory of human values (SHV) [21]. We explore the relationships between a person's values with (1) body sensors, (2) mood states, and (3) an individual's personality. Using the Happimeter system that has been developed since 2017 [8], we collected the necessary data from 2017 until now combining three channels: First, the sensor and mood data are collected through the Happimeter application using smartwatches. Second, the same application on mobile phones enables data to be transferred to the server. Third, the Happimeter website collects the value data and personality data based on the Schwartz value survey and NEO FFI test for personality [1]. The body sensors used in this research project include three categories: body movement, physiology, and context/environmental features, which are collected automatically by the Happimeter. The mood data focusses on the pleasance and activation level, which is based on users' self-report several times a day. By applying multilevel analysis to our dataset, our analysis reveals reliable support for the correlations between sensor/mood variables and values. When supplementing our framework with individuals' personality variables, we cannot find important insights based on our dataset.

The remainder of this article is organized as follows: In section "Theoretical Background", we provide a brief overview of the literature on related research. We then describe the methodology of the analysis we conducted and our findings, concluding with a discussion of our results and some future work suggestions.

Theoretical Background

Schwartz Value Theory

Many studies have utilized the Schwartz value theory. Schwartz [21] puts forth that 10 basic values, including universalism, benevolence, tradition, conformity, security, power, achievement, hedonism, stimulation, and self-direction, could be useful for understanding how people around the world think and behave. These subordinate values can be clustered into four higher order value constructs, which constitute two bipolar dimensions: openness to change versus conservation and self-enhancement versus self-transcendence [5].

The "openness to change" value dimension is defined as having autonomous thoughts and actions, and receptivity to novel experiences, while "conservation" is characterized as compliance with traditional values and customs. The first dimension captures the conflict between values that emphasize the independence of thought, action, and feelings and readiness for change and the values that emphasize order, self-restriction, preservation of the past, and resistance to change. Self-enhancement values are defined as placing importance and concern on self-interests and personal enrichment of status, while self-transcendence is operationalized as the concern for the welfare of others including those who have been marginalized. The values address egocentric desires (the pursuit of one's own interests, relative success, and dominance over others) and altruistic values (concern for the welfare and interests of others) [5, 21, 24].

Value Prediction

Figure 4.1 displays our framework, highlighting how predictors obtained from body movement combined with external influences can be applied to predict individuals' values in terms of the two bipolar dimensions: "openness to change versus conservation" and "self-enhancement versus self-transcendence". Predictors from body sensors contain three aspects: body movement, physiology, and environment feature. A second set of predictors are the mood states, which are divided into pleasance and activation. We supplement our theoretical framework with individuals' personalities, hypothesizing that they might improve the predictive quality of an individual's values.

Body Sensor and Value

A sensor generally refers to a device that converts a physical measure into a signal that is read by an observer or by an instrument. Currently, three general categories of sensors can be used for measuring physical activity in humans: movement sensors, physiological sensors, and contextual sensors [3].

Movement sensors can be used to measure human physical activities, including pedometers, gyroscopes, and accelerometers. Among these devices, accelerometers are currently the most widely used sensors for human physical activity monitoring. Physiological sensors monitor heart rate, blood pressure, temperature (skin and core body), heat flux, and so on. To date, heart rate monitoring remains the most common

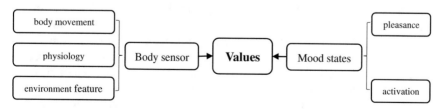

Fig. 4.1 Theoretical framework

sensor for physiological monitoring. Contextual sensors assess the context or environment in which the physical activity is being performed. Compared to motion and physiological sensors, contextual sensors are relatively new and have great potential to help describe the relationship between physical activity and various environmental features.

Using the Happimeter application, we collect data in the above-mentioned three dimensions. Bardi and Schwartz [2] demonstrated that each of the Schwartz values correlates significantly with a set of everyday behaviours. For example, power values correlate most positively with power behaviours and most negatively with benevolence behaviours. It is plausible to assume that the sensor data we collected reflects at least some causal influence of values.

We assume that people's sensors serve as predictable guides to their values related to openness to change, conservation, self-enhancement, and self-transcendence. While some sensor features are more associated with openness to alternative lifestyles and the acceptance of goals pursued by others (openness to change values), and support of justice for others (self-transcendence values), others may be more strongly influential for people who embrace authority, conformity, and traditional conceptualizations of family and society (conservation values), and pursue status and prestige (self-enhancement values) and in general are, for instance, less tolerant of homosexuality.

Mood States and Value

The second set of predictors are the mood states calculated by the Happimeter [8]. They are based on the circumplex model of affect theory, which proposes that each emotion of human beings can be understood as a linear combination of two dimensions: "valence" and "arousal" [20]. While valence is a pleasure–displeasure continuum, measuring how positive or negative an emotion is, the dimension of arousal reflects whether an emotion is exciting/agitating or calming/soothing [13]. Figure 4.2 shows the locations of different emotions which show the degree of valence and arousal each emotion presents (adopted from Lee et al. [27]). "Delighted", for example, is conceptualized as an emotional state that is associated with positive valence or pleasure together with moderate activation in the arousal dimension. Affective states other than "delighted" likewise arise from the same two dimensions but differ in the degree or extent of activation.

Emotions, by their very nature, express a personal, polarized, and biased perspective. Thus, emotion has been viewed as biasing one's evaluations, cognitions, and moral thought. The role of emotions in moral psychology has long been the focus of philosophical dispute [12]. However, all these disputes reach agreement that our mood states serve a primary role in value detection. For example, Horne and Powell [11] show that emotions are not simply experienced alongside people's judgments about moral dilemmas, but that our affective state plays a central role in determining those judgments. Eisenberg's [6] focus on guilt and sympathy shows that these higher-order emotions might motivate moral behaviour and play a role in its development

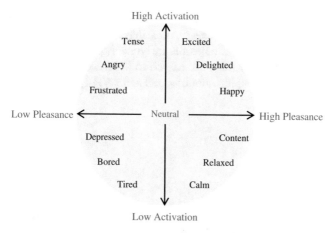

Fig. 4.2 The circumplex model of affect. *Note* The figure is adopted from Yu et al. [27]

and in moral character. Therefore, we add these two dimensions of mood states (pleasance and activation) into our experiment design. We polled users of the Happimeter system to report their levels of pleasance and activation while their sensor data was collected by the Happimeter system automatically.

Additional Tests of the Influence of Individual's Personality

Pre-tests by machine learning showed that the accuracy of Schwartz value prediction was significantly improved when users' personality variables were added into the model. In addition, there is a large extant body of literature that has explored the relationship between personality and values and verified the existence of the link between them (for instance, [7, 19, 26]. Parks-Leduc et al. [17] report a meta-analysis of 60 studies on the relations between personality traits and Schwartz values. Their findings show that openness has the most significant relationship with values. Openness correlates mostly and positively with self-direction. Moreover, openness correlates positively with stimulation and universalism, and negatively with tradition, conformity, and security. Agreeableness also has several strong associations with values, particularly and positively with benevolence. Further, agreeableness correlates positively with universalism, conformity, and tradition and negatively with power. Extraversion and conscientiousness have some moderate associations with values. However, anxiety, as a facet of neuroticism, has been associated with security [1].

Following prior literature, we used personality characteristics based on the five-factor inventory (FFI) model (neuroticism, extraversion, conscientiousness, agreeableness, and openness to experience) [4]. According to Costa and McCrae [4], neurotic people are typically distressed, depressed, impulsive, and vulnerable, and they monitor themselves closely. In turn, people characterized by openness are creative, inventive, sensitive, and open-minded. Extraverted people are social, assertive, talkative, and active, whereas those characterized by agreeableness are good-natured, compliant, and modest. Agreeable individuals are also friendly and cooperative. Finally, conscientious people are typically cautious, careful, responsible, and systematic. Personality traits are related to differences between individuals in their stable patterns of thought, emotions, and actions [14].

Data and Model

Data

The final dataset for value analysis includes 30 people who answered the Schwartz value survey at different times from 2017 to 2019; the participants include graduate students, researchers, and faculty members. Of the users who reported demographics, 37% reported their genders as male. The total number of Happimeter sensor data records for all these users is 7679. The sensor variables are directly recorded by the Happimeter application running on users' phones and smartwatches, while mood data is self-reported by users through smartwatches. Personality variables are collected through a responsive website. Only 20 of the users in our dataset could be matched with personality data. The variables list is shown in Table 4.1. Sensor and personality are continuous predictors while mood data are ordered categorical variables ranging from 0 to 2. All sensor data was standardized to facilitate interpretation of the effects.

Model

Multilevel Analysis

We use multilevel analysis to predict Schwartz values based on the sensor and mood data. The variability in the outcome can be thought of as being either within a user or between users. The data records level observations are not independent, as within a given user, data records are more similar. Figure 4.3 shows a sample where the dots are records within users, and each user is represented as a larger circle.

Mixed models incorporate fixed and random effects. A fixed effect is a parameter that does not vary, while a random effect is a parameter that varies according to the grouping variable (user), which makes it possible to explore the difference between effects within and between users. As shown in Fig. 4.4, within each user, the relation between predictor and outcome is negative. However, between users, the relation is positive. Multilevel analysis allows us to explore and understand these effects.

Table 4.1 Variables list

	Category	Variables	Definition
Values	The first dimension	Open	The sum of hedonism, stimulation, and self-direction subscores
		Conser	The sum of tradition, conformity, and security subscores
	The second dimension	Enhan	The sum of power and achievement subscores
		Trans	The sum of universalism and benevolence subscores
Sensor variables	Physiological sensors	Avgbpm	The average number of heart beats per minute
		Varbpm	The variance of heartrate per minute
	Contextual sensors	Avgnoise	The average noise level of the environment per minute
	Movement sensors	Nostep	The number of steps per minute
		Avgacc	The average of acceleration of user's movement in the physical space per minute
		Varacc	The variance of acceleration of user's movement per minute
Other variables	Mood states	Pleasance	Self-reported scores for pleasance, range from 0 to 2 (from low to high)
		Activation	Self-reported scores for activation, range from 0 to 2 (from low to high)
	FFI personality	o	Score of user's openness to experience aspect of personality
		c	Score of user's conscientiousness aspect of personality
		e	Score of user's extraversion aspect of personality
		a	Score of user's agreeableness aspect of personality
		n	Score of user's neuroticism aspect of personality

Regression Procedure

Multilevel mixed-effects generalized linear regression, using the stata mixed procedure [9, 18], was performed with 7179 data records (Level 1) across 30 individuals (Level 2) to control for the nested data structure. The models of each step are shown in Table 4.2.

Fig. 4.3 Multilevel dataset
sample

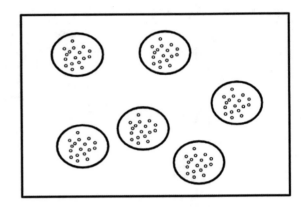

Fig. 4.4 Difference between
and within groups

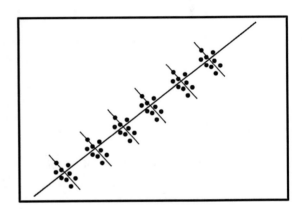

Table 4.2 Models for each step

1. Random intercept model	$V_{ij} = \gamma_{0j} + \varepsilon_{ij}$
2. Fixed sensor and mood predictors with randomly varying intercepts	$V_{ij} = \gamma_{0j} + \beta_1 S_{ij} + \beta_2 M_{ij} + \varepsilon_{ij}$
3. Fixed sensor, mood, and personality predictors with randomly varying intercepts	$V_{ij} = \gamma_{0j} + \beta_1 S_{ij} + \beta_2 M_{ij} + \beta_3 P_{ij} + \varepsilon_{ij}$

Note V = value, S = sensor predictors, M = mood predictors, P = personality predictors

Step 1 was specified as a null (baseline) model, by permitting random intercepts
only, to determine whether mean scores in different dimensions of values were significantly discrepant across all users. This model was used to compute the intraclass
correlation (ICC), an indication of the extent that sensor data of the same user were
similar on their value scores relative to the total variation in sensor data among all
users. A high ICC value beyond the null hypothesis of 0.00 signifies that sensor data
units are not statistically independent within a certain user, and therefore the nested
design should be considered by using a multilevel model.

Step 2 involved random intercepts with fixed-effect predictors. It builds on the previous model by including the fixed-effect predictors at the data records level (sensor variables and mood variables). Thus, Step 2 controlled for the nested structure by permitting intercepts to vary, while estimating fixed effects of the relevant variables.

Building on Step 2, Step 3 incorporated the user-level personality variables. They were added into the models to test the relationships between personality variables and Schwartz values.

Results

Descriptive information for all variables is presented in Table 4.3. Following the manual of the Schwartz value survey [23], we centred the score of each questions by the average score of each user. Then the four values were calculated based on the centre-scored results of all 10 value questions. From Table 4.3 we see that the mean value of openness is larger than that of conservation, while the averages of self-transcendence are larger than self-enhancement, which means that in our dataset

Table 4.3 Descriptive analysis of variables

Variable	Obs	Mean	Std. Dev.	Min	Max
Open	7,679	0.928	1.583	−4.9	8.1
Conser	7,679	−1.621	2.396	−7.2	4
Enhan	7,679	−1.574	2.843	−7	4.4
Trans	7,679	2.267	2.412	−2.6	6.8
Pleasance	7,679	1.463	0.555	0	2
Activation	7,679	1.183	0.58	0	2
Std avgbpm	7,679	0	1	−2.09	4.876
Std varbpm	7,679	0	1	−1.083	8.737
Std nostep	7,679	0	1	−0.459	7.788
Std avgnoise	7,679	0	1	−0.883	9.885
Std varnoise	7,679	0	1	−0.697	8.74
Std avgacc	7,679	0	1	−4.926	5.082
Std varacc	7,679	0	1	−1.839	6.95
o	6,186	0.585	0.0829	0.375	0.733
c	6,186	0.685	0.0884	0.521	0.767
e	6,186	0.684	0.0614	0.55	0.783
a	6,186	0.582	0.0678	0.375	0.683
n	6,186	0.471	0.0913	0.271	0.683
Gender	7,679	0.324	0.468	0	1

Table 4.4 Intraclass correlation of null models

Values	ICC	Std. Err.	95% Conf.	Interval
Open	0.946	0.0147	0.908	0.968
Conser	0.735	0.0604	0.602	0.836
Enhan	0.531	0.0791	0.378	0.679
Trans	0.617	0.0745	0.465	0.749

people tend to regard themselves as open to change and self-transcendent instead of conservative or self-enhancing. The correlation matrix of all predictor variables is presented in Table 4.5. It reveals that multi-collinearity exists between different indexes of personality, which is taken into consideration for the regression analysis. In addition, the correlations between avgacc and varacc, avgnoise, and varnoise are also higher than the rule of thumb (0.7), thus we removed varacc and varnoise from our final models.

Random Intercept Model

We hypothesized that characteristics attributed to the user level would explain variation in values. Based on the null model, the results (Table 4.4) reveal significant ICCs 94.6, 73.5, 53.1, and 61.7% for all dimensions, signifying that over 50% of the variance in one's value is explained exclusively by variations across users. This provided sufficient evidence that a multilevel regression model was warranted [9, 18].

Fixed Sensor and Mood Predictors with Randomly Varying Intercepts

Models 1–4 in Table 4.6 tested the fixed-effect for sensor and mood predictors at level 1. The results show that:

(1) Sensor-level variables are significantly related to the four aspects of the Schwartz values. Specifically, the average of heartrate is positively associated with conservation and self-transcendence, while negatively related to openness, which indicates that people who are open to change and who focus on self-development tend to have a relatively lower heart beat than people who are conservative. Regarding the variance of heartrate, we note that self-enhancing individuals tend to have low heartrate variability. For activity-related variables, neither the number of steps nor the average of acceleration is correlated with the Schwartz values of users; however, the standard deviation of all activity-related variables

is correlated with most Schwartz values; in general, the higher the standard deviation, the more open and the less conservative people are (Table 4.5). Regarding environmental attributes, for the noise level we found that people who are open to change and focus on transcendence are more likely to be in a quieter environment, whereas those who are conservative and pay attention to self-development seem to be in noisier environments.

(2) Mood variables are also related to the values of people. We find that pleasance and activation vary in the way they relate to the Schwartz values. Figure 4.5 shows the spectrum of Schwartz values for a person across our sample users. Open and self-enhancing people have higher tendency for pleasance but lower activation. This is somewhat surprising, as we commonly tend to regard self-transcendent people as happy and satisfied. It could be that in our sample self-transcendent people are more critical and questioning against themselves, which might reduce their happiness at times.

Looking at gender, we find that in our sample women tend to be less open and more conservative than men (gender has been coded as male = 1, female = 0).

Table 4.6 includes the results of the regressions for the four Schwartz values using fixed sensor, mood, and personality predictors with randomly varying intercepts.

Using FFI Personality as Additional Predictors or Moderating Variables

As the correlation matrix in Table 4.5 shows, high relative coefficients exist among the five personality variables. Taking this into consideration, we conducted further analysis using different methods to test the relationships between FFI personality and Schwartz values.

First, we add agreeableness, neuroticism, extraversion, and conscientiousness into our models while removing the openness personality variable. According to Table 4.5, severe multi collinearity only exists between the personality variable openness and other personality variables (with agreeableness 0.84, neuroticism 0.70, agreeableness 0.69, and extraversion 0.50). After removing the openness variable, none of the other correlated coefficients is higher than the threshold of 0.7, which is used as a rule of thumb in literature. However, the models with dependent variables in the first dimension of value (openness to change and conservation) do not concave when adding the four personality variables to the models. For the second dimension (self-enhancement and self-transcendence), including the personality characteristics into the regression also does not lead to reliable results. Encouraged by existing studies (i.e., [10]) we were looking for a better fit by adding one personality variable into the models at a time to avoid the multi-collinearity problems. Unfortunately, that did not work with our dataset neither. Finally, we also unsuccessfully tested indirect or moderating effect of users' personality on the Schwartz values. In conclusion, no solid evidence was found with the above methods to support the relationship between FFI personality and Schwartz values based on our dataset.

Table 4.5 Correlation matrix

		1	2	3	4	5	6	7	8	9	10	11	12	13	14	15	16	17	18	19
1	Open	1																		
2	Conser	−0.50*	1																	
3	Enhan	−0.28*	−0.31*	1																
4	Trans	0.17*	−0.30*	−0.69*	1															
5	Pleasance	0.04*	0	0.03*	−0.05*	1														
6	Activation	−0.03*	0.01	−0.06*	0.08*	0.26*	1													
7	Std avgbpm	−0.01	0.11*	−0.08*	−0.01	−0.01	0.04*	1												
8	Std varbpm	0.09*	−0.01	−0.04*	0	0.04*	0.01	0.12*	1											
9	Std nostep	0.14*	−0.16*	0.05*	0.01	−0.03*	−0.04*	0.22*	0.14*	1										
10	Std avgnoise	−0.05*	0.16*	0.07*	−0.20*	0.06*	−0.03*	0.19*	−0.13*	−0.02	1									
11	Std varnoise	0.01	0.17*	−0.06*	−0.11*	0.02	0	0.10*	−0.08*	−0.04*	0.53*	1								
12	Std avgacc	−0.08*	0	0.04*	0	−0.01	−0.04*	−0.05*	−0.15*	0.03*	0.07*	−0.06*	1							
13	Std varacc	0.10*	−0.05*	−0.02	0.01	0.02	0.05*	0.36*	0.16*	0.28*	0.12*	0.09*	−0.67*	1						
14	n	−0.38*	0.41*	0.21*	−0.46*	0.09*	0.09*	0.03*	−0.09*	−0.25*	0.18*	0.11*	−0.03*	−0.03*	1					
15	e	−0.14*	0.54*	−0.34*	−0.03*	0.02	0.09*	0.12*	0.11*	−0.03*	−0.05*	0.04*	−0.07*	0.07*	0.45*	1				
16	o	−0.15*	0.63*	−0.34*	−0.12*	0.07*	0.13*	0.11*	−0.02	−0.30*	0.16*	0.18*	−0.04*	−0.03*	0.70*	0.50*	1			
17	a	−0.34*	0.64*	0.04*	−0.49*	0.07*	0.05*	0.03*	0.01	−0.23*	0.12*	0.17*	0	−0.07*	0.51*	0.17*	0.69*	1		
18	c	−0.15*	0.55*	−0.11*	−0.32*	0.08*	0.09*	0.11*	0.09*	−0.18*	0.14*	0.13*	−0.04*	0.01	0.62*	0.62*	0.84*	0.58*	1	
19	Gender	0.19*	−0.54*	0.20*	0.18*	−0.01	0.06*	0.01	0.17*	0.17*	−0.17*	−0.13*	0	0.04*	−0.46*	−0.15*	−0.28*	−0.27*	0.04*	1

*p < 0.1

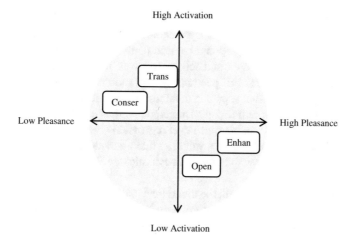

High Activation

Trans

Conser

Low Pleasance

Enhan

Open

High Pleasance

Low Activation

Fig. 4.5 Relationships of mood states and values

Table 4.6 Regression results

	Model 1	Model 2	Model 3	Model 4
Variables	Open	Conser	Enhan	Trans
Avgbpm	−0.014**	0.261***	−0.236***	0.003
Varbpm	0.068***	0.112***	−0.252***	−0.013
Nostep	−0.000	−0.233***	0.062**	0.054**
Avgacc	−0.016**	−0.027	0.016	0.028
Avgnoise	−0.085***	0.241***	0.445***	−0.536***
Pleasance	0.028**	−0.112***	0.180***	−0.138***
Activation	−0.045***	0.154***	−0.343***	0.317***
Gender	0.841	−2.660***	1.575***	0.515***
Constant	0.733***	−0.733***	−1.773***	1.742***
Observations	7,679	7,679	7,679	7,679
Number of groups	30	30	30	30

Note Standard errors in parentheses ***$p < 0.01$, **$p < 0.05$, *$p < 0.1$

Discussion

The ethical values of individuals have captured the interest of researchers, practitioners, social critics, and the public at large [15]. Schwartz [22] defined values as reflected in the course of action in an individual's life. However, prior literature mainly focuses on measuring people's values via surveys. The advent of sensing technology provides a powerful solution to the challenge of detecting an individual's values. Smartwatch sensors provide a simple way of passively detecting the body signals and the environment a user is encountering, also reducing the burden of self-reporting. Through our unique Happimeter mood sensing system, we were able to gather data about body signals and environmental features sensed by the smartwatches, self-reported pleasance and activation levels, in combination with personal data about Schwartz values, FFI personality, and morals entered through a survey on a website. We found that a person's values are reflected by their body sensors and mood states. What's more, body language also has a strong relationship with a person's personality. By using multilevel regressions, we showed a link between sensor/mood variables and people's values, while no evidence was found to show a moderating effect of FFI personality characteristics on Schwartz values, at least in our dataset.

This article contributes to both the theoretical and practical sides. With respect to furthering the state of research literature, we see our study having two contributions. First, we provide a novel method of detecting people's ethical values based on sensing technologies, going beyond traditional survey methods. Prior research acquires the value of an individual mainly by questionnaire, and this study fills the academic gap and indicates the potential of applying a sensing system for value detection. Second, we propose an integrative framework of using body signals, mood states, and environment features to study ethical values. We highlighted that the body signals that a person displays, the environment that people live in, and the mood states of people may be consistently associated with their perceptions and behaviours, and thus have psychological implications and clear links to a person's values.

Our findings also provide important insights for the real world. First, by applying sensing technology, individuals become more aware of their values, which helps them to make choices with which they feel comfortable. They can also become more aware of how they come across to others. Self-awareness is a first step in any change. They would know if they need to make moves based on their image of themselves. Second, pairs working in a team that might have had difficulties working well together can better identify the cause of blockages or conflict, which might be related to values. The better knowledge of individuals' values can help us design better teams and manage the relationships in teams more effectively.

However, our study inevitably has some limitations. First, this study only focusses on a set of limited variables (mainly heartrate, acceleration, noise level, and pleasance). As technology continues to develop and research continues to identify new predictors that are psychologically meaningful, future work will be able to investigate the collective and interactive effects of these additional factors on people's values. For instance, researchers could integrate stress level, light level, and other relevant variables into their models. Second, the combination of mood states and smartwatch sensing allowed us to collect large amounts of within-person data in the current work. However, the number of participants in our dataset is limited. Further analysis will have to be done with larger numbers of participants.

In sum, we have identified novel links between body postures and body language, emotions, and ethical values, showing that how one behaves really tells who he/she is.

References

1. A. Aluja, L.F. Garcia, Relationships between big five personality factors and values. Soc. Behav. Personal. Int. J. **32**(7), 619–625 (2004)
2. A. Bardi, S.H. Schwartz, Values and behavior: Strength and structure of relations. Pers. Soc. Psychol. Bull. **29**, 1207–1220 (2003)
3. K.Y. Chen, K.F. Janz, W. Zhu, R.J. Brychta, Re-defining the roles of sensors in objective physical activity monitoring. Med. Sci. Sports Exerc. **44**(1 Suppl 1), S13 (2012)
4. P. Costa, R.R. McCrae, A theoretical context for adult temperament. Temper. Cont. 1–21 (2001)
5. E. Davidov, P. Schmidt, S.H. Schwartz, Bringing values back in: the adequacy of the European Social Survey to measure values in 20 countries. Pub. Opin. Quart. **72**(3), 420–445 (2008)
6. N. Eisenberg, Emotion, regulation, and moral development. Annu. Rev. Psychol. **51**(1), 665–697 (2000)
7. R. Fischer, D. Boer, Motivational basis of personality traits: a meta-analysis of value-personality correlations. J. Pers. **83**(5), 491–510 (2015)
8. P.A. Gloor, A.F. Colladon, F. Grippa, P. Budner, J. Eirich, Aristotle said "Happiness is a State of Activity" — predicting mood through body sensing with smartwatches. J. Syst. Sci. Syst. (2018)
9. R.H. Heck, S.L. Thomas, L.N. Tabata, Multilevel and longitudinal modeling with IBM SPSS (2nd ed.) (Routledge, New York, NY, 2013)
10. M. Hietalahti, A. Tolvanen, L. Pulkkinen, K. Kokko, Relationships between personality traits and values in middle aged men and women. J. Happi. Well-Being **6** (2018)
11. Z. Horne, D. Powell, How large is the role of emotion in judgments of moral dilemmas? PLoS ONE **11**(7), e0154780 (2016)
12. B. Huebner, S. Dwyer, M. Hauser, The role of emotion in moral psychology. Trends Cognit. Sci. **13**(1), 1–6 (2009)
13. E.A. Kensinger, Remembering emotional experiences: the contribution of valence and arousal. Rev. Neurosci. **15**(4), 241–252 (2004)
14. R.R. McCrae, P.T. Jr. Costa, Personality in adulthood. A five-factor theory perspective. (2nd ed.) (The Guilford Press, New York, 2003)
15. B.M. Meglino, E.C. Ravlin, Individual values in organizations: concepts, controversies, and research. J. Manag. **24**(3), 351–389 (1998)
16. D.J. Ozer, V. Benet-Martinez, Personality and the prediction of consequential outcomes. Annu. Rev. Psychol. **57**, 401–421 (2006)
17. L. Parks-Leduc, G. Feldman, A. Bardi, Personality traits and personal values: a meta-analysis. Pers. Soc. Psychol. Rev. **19**, 3–29 (2015)
18. J. L. Peugh, C. K. Enders, Using the SPSS mixed procedure to fit cross-sectional and longitudinal multilevel models. Educ. Psychol. Meas. **65**, 717–741 (2005)
19. S. Roccas, L. Sagiv, S.H. Schwartz, A. Knafo, The big five personality factors and personal values. Pers. Soc. Psychol. Bull. **28**(6), 789–801 (2002)
20. J.A. Russell, A circumplex model of affect. J. Pers. Soc. Psychol. **39**(6), 1161 (1980)
21. S.H. Schwartz, Universals in the content and structure of values: theoretical advances and empirical tests in 20 countries. Adv. Experi. Soc. Psychol., Elsevier. **25**, 1–65 (1992)
22. S. H. Schwartz, Are there universal aspects in the content and structure of values? J. Soc. Issues **50**, 19–46 (1994)
23. S.H. Schwartz, Draft users manual: proper use of the Schwarz value survey, version 14 January 2009, compiled by R.F. Littrell (Auckland, Centre for Cross Cultural Comparisons, New Zealand, 2009)
24. S.H. Schwartz, G.V. Caprara, M. Vecchione, Basic personal values, core political values, and voting: a longitudinal analysis. Politi. Psychol. **31**(3), 421–452 (2010)
25. S.H. Schwartz, T. Butenko, Values and behavior: validating the refined value theory in Russia. Europ. J. Soc. Psychol. **44**(7), 799–813 (2014)

26. M. Vecchione, G. Alessandri, C. Barbaranelli, G. Caprara, Higher-order factors of the big five and basic values: empirical and theoretical relations. Br. J. Psychol. **102**(3), 478–498 (2011)
27. L.-C. Yu, L.-H. Lee, S. Hao, J. Wang, Y. He, J. Hu, K. Lai, X. Zhang, Building Chinese affective resources in valence-arousal dimensions (2016)

Chapter 5
Measuring Workload and Performance of Surgeons Using Body Sensors of Smartwatches

Juan A. Sánchez-Margallo, Peter A. Gloor, José L. Campos, and Francisco M. Sánchez-Margallo

Abstract We present the first steps toward building an intelligent system to measure the workload and surgical performance of minimally invasive surgeons. This pilot study was conducted during two training courses in minimally invasive suturing, one in microsurgery and one in laparoscopic surgery. During each training activity, surgeons wore a smartwatch with the Happimeter application running on it. This system recorded a set of physiological and motion parameters during the surgical execution. We found that monitoring the surgeon's maneuvers and physiological parameters during surgical activity has the potential to play an important role in predicting the workload and surgical performance, especially regarding physical and mental demand and the level of distraction during surgery.

Introduction

The introduction of laparoscopic surgery has made it possible to reduce the number of incisions required during the surgical procedure, minimizing the trauma and therefore reducing the associated pain, the risk of infection, and the length of hospital stay. In laparoscopy, surgeons operate through surgical ports on the patient's abdominal wall and using long instrument and endoscopic camera. In the case of microsurgery,

J. A. Sánchez-Margallo (✉) · J. L. Campos · F. M. Sánchez-Margallo
Jesús Usón Minimally Invasive Surgery Centre, Cáceres, Spain
e-mail: jasanchez@ccmijesususon.com

J. L. Campos
e-mail: jlcampos@ccmijesususon.com

F. M. Sánchez-Margallo
e-mail: msanchez@ccmijesususon.com

P. A. Gloor
MIT Center for Collective Intelligence, Cambridge, MA, USA
e-mail: pgloor@mit.edu

© Springer Nature Switzerland AG 2020
A. Przegalinska et al. (eds.), *Digital Transformation of Collaboration*,
Springer Proceedings in Complexity,
https://doi.org/10.1007/978-3-030-48993-9_5

67

surgeries are performed by means of visual magnification and microsurgical instruments to carry out interventions on the vascular and nervous systems, among others. Apart from the numerous benefits for the patient, these surgical techniques present several limitations for the surgeon. Some of these challenges are loss of depth perception (two-dimensional vision), forced postures during surgery due to the restricted movements, diminished tactile feedback, and inverted instrument movements due to the surgical ports. All of this leads to an increase in the surgeon's mental and physical workload during surgery, as well as a potential onset of musculoskeletal disorders.

Several wearable devices have been recently developed for surgical applications, and most of them focused on interaction with patient's preoperative information and telementoring purposes [1, 2]. However, there is a lack of this technology that allows us to monitor and analyze the wellbeing of the surgeon while operating, as well as the development of the minimally invasive surgical procedures. To the best of our knowledge, this is the first project focused on the use of wearable technology and artificial intelligence to look for solutions in the prevention of emerging health problems of surgeons, as well as in the prediction of the quality of his/her surgical performance during the surgical practice. In this work, we present the first steps toward building an intelligent system for measuring the workload and surgical performance of minimally invasive surgeons.

Materials and Methods

This pilot study was conducted during two training courses in minimally invasive suturing. One course was focused on microsurgical techniques and the other on laparoscopic surgery. During each training activity, surgeons wore a smartwatch with the Happimeter app running on it. This system [3] recorded a set of user motion and physiological parameters during the surgical performance. The smartwatch collects body movements through the accelerometer sensor of the smartwatch, heartrate through the heartrate sensor, speech parameters (no content) through the microphone, close interaction through the Bluetooth sensor, and location changes through the GPS. The data is transmitted from the watch (currently we are using the Android Wear Ticwatch) to a server in the cloud, where a machine learning system collects the data, trains the models, and predicts pleasance, activation, and stress levels. With regard to the motion parameters, the values (direction) of the X, Y, and Z components of the acceleration in the movements of the surgeon's hand (AccelerometerX, AccelerometerY, and AccelerometerZ) and their magnitudes (i.e. change in velocity; AccelerometerMagX, AccelerometerMagY, and AccelerometerMagZ) were used.

At the end of each trial, participants were asked to complete the Surgery Task Load Index (SURG-TLX) [4] questionnaire, which is a subjective questionnaire to evaluate the workload during a surgical activity. This multidimensional assessment method is based on six dimensions defined as mental demand, physical demand, temporal demand, task complexity, situational stress, and distractions.

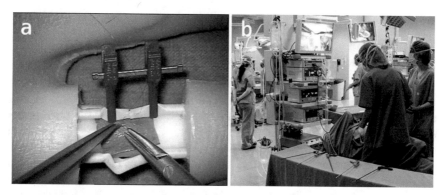

Fig. 1 Microsurgical anastomosis on a simulator (**a**). Training course on laparoscopic surgery (**b**)

Microsurgical Training

The first training activity in this study was a course in nerve and vascular microsurgery. Resident surgeons, with different experience levels, performed a vascular anastomosis using a physical simulator (Fig. 1a).

During this activity, the surgeons' technical skills were assessed using the Stanford Microsurgery and Resident Training (SMaRT) scale [5]. The SMaRT scale consists of nine categories graded on a 5-point Likert scale, including instrument handling, respect for tissue, efficiency, suture handling, suturing technique, quality of knot, final product, operation flow, and overall performance.

Laparoscopic Training

The second activity was a course on laparoscopic suture, in which surgeons, with different levels of experience, were evaluated during the performance of a laparoscopic gastrostomy in a porcine model (Fig. 1b). During the course of the training activity, participants exchanged the roles of principal surgeon and camera assistant. Therefore, in this case, the quality of surgical performance was not individually assessed.

All surgical procedures were reviewed and approved by the competent local Animal Welfare and Ethics Committee. Accommodation of the animals and their handling were done in accordance with the European directive (2010/63/EU), Spanish laws (RD 53/2013) for animal use and care, ARRIVE guidelines, and according to the Guide for the Care and Use of Laboratory Animals.

Results

Microsurgical Training

Eight resident surgeons, between first and fifth year of residency and an average experience of less than 10 microsurgical procedures performed, participated in this study. With regard to the SMaRT assessment, the surgical skills of the residents during the performance of the microsurgical tasks were generally scored as moderate, specifically concerning the respect of tissue, the quality of the final product, and the overall performance (Fig. 2). All participants completed the surgical workload questionnaire. In general, they scored the training task as physically demanding and complex.

Regarding the correlation between the motion and physiological parameters and the surgical skills during the microsurgical activity, there was a positive correlation between the magnitude of the Y component of the hand acceleration and the quality of the suturing technique and knot tying (Fig. 3). Concerning the surgical workload, there were correlations between the acceleration of hand movements in the X axis and the mental demand, complexity of the task, and level of distraction during the task performance (Fig. 4). Additionally, there were also strong correlations between the acceleration of the hand motion in the Z axis and the temporal demand of the task and between the pitch/tone of the voice and both the distractions during the task and the surgeon's heart rate.

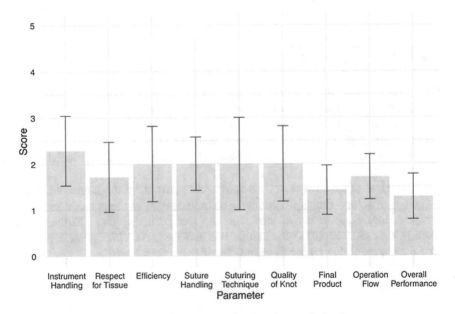

Fig. 2 Average values of the SMaRT parameters for the microsurgical task

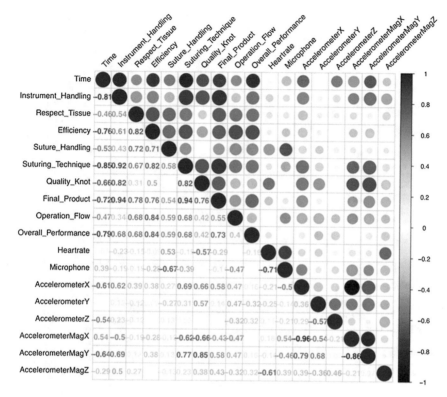

Fig. 3 Correlation matrix between the Happimeter parameters and the SMaRT categories for the microsurgical task. The correlation coefficients are shown at the lower part of the graph. Color intensity and the size of the circle are proportional to the correlation coefficients

Laparoscopic Procedure

During the laparoscopic course, seven surgeons participated in this study. They were one third-year, one fourth-year, and two fifth-year resident surgeons and three consultant surgeons, with an average experience of between 10 and 50 laparoscopic procedures performed. All participants completed the surgical workload questionnaire. They reported high values of mental, physical, and temporal demands and stress during the execution of the laparoscopic procedure. They also reported that the level of complexity of the task was high. Results showed strong correlations between the acceleration of the hand movements, mainly for the X and Y components, and the level of physical demand during the laparoscopic procedure (Fig. 5).

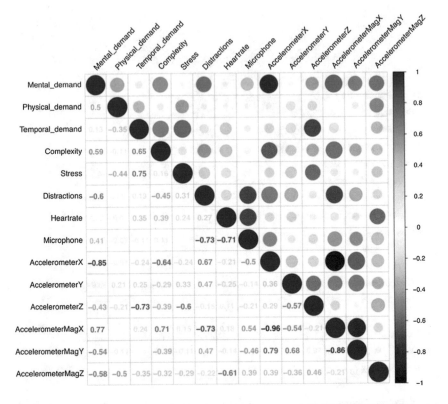

Fig. 4 Correlation matrix between the Happimeter parameters and the SURG-TLX factors for the microsurgical task

Discussion

Participants in this study considered the system an attractive solution to assist the surgical practice, seeking to improve the surgeon's physiological conditions and surgical performance during surgical practice. They stated that the device did not bother them during the course of the training activity or in the performance of the surgical tasks and procedures. In addition, the system does not require personal data of the subject other than information related to hand movements, heart rate, and tone of voice. Participants were well-disposed to provide feedback and complete the questionnaire regarding their workload levels during the course of the study. Providing information of the surgeon's mental and physical workload, distraction and stress levels, and quality of the surgical technical performance can have a significant impact on his or her surgical outcome.

In general, surgeons considered both microsurgical tasks and laparoscopic procedures to be physically demanding. In both disciplines of minimally invasive surgery,

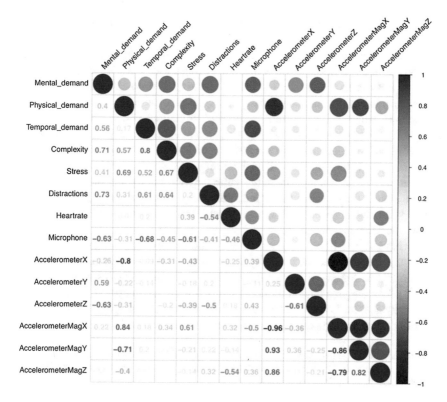

Fig. 5 Correlation matrix between the Happimeter parameters and the SURG-TLX factors for the laparoscopic procedure

the surgeon's hand movements seem to be quite related to the physical and mental demand of the surgical task or procedure being performed.

During the microsurgical course, the surgeons' surgical performance was generally scored as moderate. This could be due to their relatively low experience in microsurgery and to the fact that the task evaluated was one of the first in the entire microsurgical training course.

Analyzing the relationships between the motion and physiological parameters recorded by the Happimeter system and the microsurgical performance, it seems that the magnitude of the X and Y components of the hand acceleration during the microsurgical performance could be used as a part of a further model for predicting the quality of microsurgical suture. Additionally, the acceleration of the X component may have a strong relationship with how mentally demanding and complex a microsurgical task is considered by a novice surgeon. Therefore, these factors could be used as indicators of surgical competence and how confident or stressed a novice surgeon feels during the performance of a task. A strong correlation was also found between the tone of voice/noise during the execution of the surgical activity and the distraction levels and the heart rate of the surgeon. This environmental parameter may be considered a potential factor influencing the concentration and stress levels of the surgeon.

In the case of laparoscopic procedure, there was a strong correlation between hand motion and the increased physical demand during surgery. As we might expect, an increase in hand movements using the laparoscopic instruments generally leads to an increase in physical demand during a surgical procedure.

This preliminary study presents a series of limitations which will be considered in future studies. The surgeons who participated in the laparoscopy course alternated in their roles as surgeon and camera assistant on several occasions during the course of the activity. Future studies will analyze surgical tasks or procedures in which the surgeon and the assistant perform the same function throughout the procedure and thus be able to fully evaluate their individual surgical performance. The number of participants and surgical tasks were limited. Further studies with a larger sample should therefore be carried out.

We have found that tracking surgeons' motion and physiological parameters during the surgical practice may play an important role in predicting their workload and surgical performance. Taking into account the results of both studies, the acceleration of the surgeon's hand motion, mainly for the X component, may be related to the physical and mental demand and complexity of a surgical task (surgeon's workload), and the tone of voice/noise to the level of distraction during surgery.

Acknowledgements This work has been partially funded by the MISTI Global Seed Funds, "la Caixa" Foundation (LCF/PR/MIT18/11830006), Junta de Extremadura (Spain), and European Regional Development Fund (GR18199).

References

1. F.M. Sánchez-Margallo, J.A. Sánchez-Margallo, J.L. Moyano-Cuevas, E.M. Pérez, J. Maestre, Use of natural user interfaces for image navigation during laparoscopic surgery: initial experience. Minim. Invasive Ther. Allied Technol. **26**, 253–261 (2017)
2. H.A.W. Meijer, J.A. Sánchez Margallo, F.M. Sánchez Margallo, J.C. Goslings, M.P. Schijven, Wearable technology in an international telementoring setting during surgery: a feasibility study. BMJ Innov. **3**, 189–195 (2017)
3. P.A. Gloor, A.F. Colladon, F. Grippa, P. Budner, J. Eirich, Aristotle Said "Happiness is a State of Activity"—predicting mood through body sensing with Smartwatches. J. Syst. Sci. Syst. Eng. **27**(5), 586–612 (2018)
4. M.R. Wilson, J.M. Poolton, N. Malhotra, K. Ngo, E. Bright, R.S.W. Masters, Development and validation of a surgical workload measure: the surgery task load index (SURG-TLX). World J. Surg. **35**, 1961–1969 (2011)
5. T. Satterwhite, J. Son, J. Carey, A. Echo, T. Spurling, J. Paro, G. Gurtner, J. Chang, G.K. Lee, The Stanford Microsurgery and Resident Training (SMaRT) scale. Ann. Plast. Surg. **72**, S84–S88 (2014)

Chapter 6
Exploring the Impact of Environmental and Human Factors on Operational Performance of a Logistics Hub

Davide Aloini, Giulia Benvenuti, Riccardo Dulmin, Peter A. Gloor, Emanuele Guerrazzi, Valeria Mininno, and Alessandro Stefanini

Abstract This work aims to explore the environmental and human factors affecting productivity of warehouse operators in material handling activities. The study was carried out in a semi-automated logistic hub and the data collection has been conducted using wearable sensors able to detect human-related variables such as heart rate and human interactions, based on a smartwatch combined with a mobile application developed by the MIT Center for Collective Intelligence. Preliminary analysis has shown that the interaction between the warehouse operators and the team leader significantly affects the productivity.

Introduction

The importance of warehousing as a key activity for manufacturing companies is widely recognized by scholars and practitioners. Although in recent years a more pervasive automation under the label "Industry 4.0" is emerging, a large number of warehouses still strongly rely on human workers, such as forklift drivers, especially in picking activities. Indeed, the picking process is considered among the most laborious and expensive activities in warehouses, and estimates suggest that around 80% of the order picking is still carried out manually [1].

In such a context, where individuals are predominantly involved in material handling activities, individual and team dynamics are among the fundamental drivers of workplace motivation, wellness and system productivity [2]. As a result, there is a growing interest in taking into consideration human factors such as stress, health and

D. Aloini · G. Benvenuti · R. Dulmin · E. Guerrazzi (✉) · V. Mininno · A. Stefanini
University of Pisa, Largo Lucio Lazzarino, 56122 Pisa, Italy
e-mail: Emanuele.guerrazzi@phd.unipi.it

D. Aloini
e-mail: davide.aloini@unipi.it

P. A. Gloor
Massachusetts Institute of Technology, 77 Massachusetts Avenue, Cambridge, MA, USA

© Springer Nature Switzerland AG 2020
A. Przegalinska et al. (eds.), *Digital Transformation of Collaboration*,
Springer Proceedings in Complexity,
https://doi.org/10.1007/978-3-030-48993-9_6

collaboration dynamics in the logistic operations. Nevertheless, while such factors promise to achieve better business performance and high levels of safety and well-being of workers, there is still a lack of studies [3] that consider both human and working environment factors in such a context. Indeed, studying these variables quantitatively can lead to a clearer insight on which are the factors that affect productivity in such logistics operations.

The recent advent of wearable sensors, and other sensor-based measurement tools (e.g. sociometric badges and smartwatches), is offering researchers the opportunity of collecting and analyzing human factors through data-driven methodologies [4]. For instance, smartwatches are particularly suitable to measure acceleration and heart rate, useful as predictors for physical fatigue of workers [5], and they can efficiently collect a big amount of data in real time, avoiding the main bias linked to common approaches in behavioral studies and increasing the data richness, quality and reliability. On the other hand, more conventional instruments such as the thermo-hygrometer and lux-meter can be used to measure data about the quality of the environment (e.g. temperature, humidity and luminosity).

Background

Many researchers developed analytical models to improve picking activities and increase efficiency, suggesting different warehouse layouts, routes or archiving tasks, but only a minority of them focused their studies on the human factor, as highlighted in the study of Grosse et al. [3] and reported in Fig. 6.1.

This study suggests that future research should focus on the connection between the order picking system design and the aspects of human factors and it indicates opportunities of integrating aspects of human factors in management-oriented warehouse picking research. A more detailed study about human factors and ergonomics could be promising to achieve higher performance (time/productivity) and quality (reduced errors) and to improve safety and well-being of workers, as suggested by Neumann et al. [6].

Moreover, a logistics hub is a complex environment where external factors can affect performance, such as the carrier selection [7] or fragmentation and scarce collaboration among the extensive networks of suppliers and subcontractors [8]. Regarding human factors, there is a multitude of risks and dangers regarding the working environment that leads research conduct studies about environmental factors such as temperature, humidity and luminosity, as reported by Leather et al. [9].

Our goal is to embrace all these factors together and to collect quantitative data in order to explore possible connections among them.

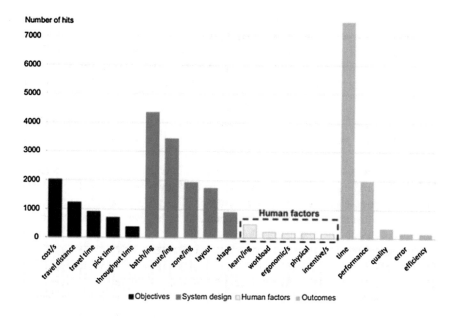

Fig. 6.1 Literature studies about human factors in warehousing (*Source* [3])

Research Objective and Methodology

This research aims to empirically explore environmental and human factors that may affect productivity and more generally performance of operators in material handling activities. Toward this goal, an exploratory case study was conducted in a semi-automated logistics hub, located in Tuscany, Italy.

The research methodology includes the following steps:

(1) *Context analysis.* A context analysis of the logistics hub has been done for setting the case up and to test preliminarily the innovative measurement tools in the specific application context (e.g. smartwatches, thermo-hygrometer and lux-meter with datalogger function).

(2) *Data collection.* Data have been collected from employees through the smartwatches (e.g. body movement, acceleration, proximity, speaking, heart rate), while we used thermo-hygrometer and lux-meter for working environmental measures. Finally, productivity data are provided by the warehouse management systems (WMS).

(3) *Data pre-processing.* Data from different sources have been properly preprocessed and homogenized to obtain comparable data useful for the analysis phase. Moreover, in order to avoid bias due to the physiological differences among workers (i.e. different heart rate baseline), smartwatch data have been normalized for each operator. Then, additional features were created through

the combination of the measured variables, for example, the mood variables, based on Russell's circumplex model [10].

(4) *Data analysis*. Correlation and regression analyses have been conducted to propose an explanatory model that can provide guidelines for management with directions for possibly improving workers' productivity, safety and well-being.

Case Study

The experiment has been carried out in a semi-automated logistics hub, located in Tuscany, Italy. The hub is owned by a 450+ million revenue company that produces and sells paper-tissue products all over Europe. The warehouse covers an area of 24,000 m^2 with a height of 7 m and it is open 24 h a day from Monday to Saturday afternoon. Operators work in three shifts.

Operational activities carried out in the warehouse consist of the stocking of inbound pallets and the picking of ordered pallets that must be delivered. Each pallet is 80 cm × 120 cm and the weight is not considered critical due to the lightness of the paper. About 10 operators are employed for each shift, five operators for picking, four for stocking, one team leader, and people from administrative staff.

The layout of the warehouse is shown in Fig. 6.2. It is divided into six areas. Each of the zones numbered from 1 to 5 has one pallet shuttle, while zone 6 is managed using stacks, without any automation. The use of pallet shuttles provides more density of shelves in the warehouse, as well as a greater safety for the operators.

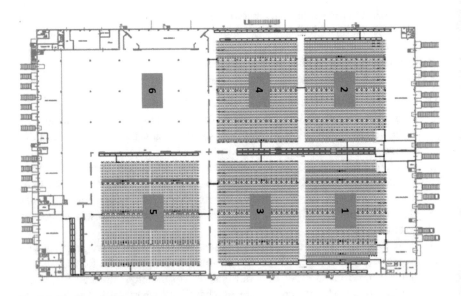

Fig. 6.2 Logistics hub layout

We note that although the pallet shuttle partially supports warehouse automation, the presence of workers is still very necessary.

Data Collection

Human and environmental data have been collected from the warehouse 8 h a day, 5 days a week, 8 weeks from March to May 2019, for a total of 320 h monitored for each operator. Figure 6.3 provides an overview of the hardware and software used for data collection.

As for human factors, a group of five picking workers, including the team leader, has been equipped with smartwatches and monitored during working activities for the whole duration of the experiment. The tool allows researchers to conduct more in-depth quantitative analysis of interactions inside organizations leading to high-level descriptions of human behavior in terms of (i) physical activity/human movement, (ii) speech features (rather than raw audio), (iii) indoor localization and (iv) proximity to other individuals.

As concerning environmental variables, a thermo-hygrometer and a lux-meter, both with data logger function, have been positioned near to the outbound area of the warehouse for collecting environmental data.

Finally, productivity is evaluated through the hourly number of processed orders (data were extracted by the WMS).

Figure 6.3 summarizes our data collection instrumentation.

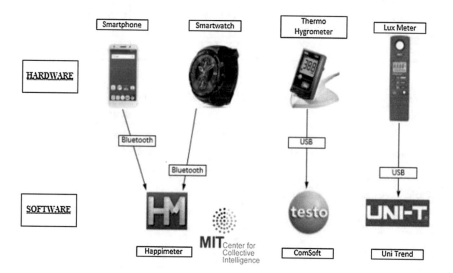

Fig. 6.3 Hardware and software for data collection

Results

The most significant correlations and p-values are presented in Table 6.1. Results show that there is a negative correlation with "worked_hours", that is, the number of hours worked by an operator, while the correlation with "team leader connection", that is, the measurement of how much an operator gets in touch with the team leader, is positive. As an addition, as concerning the features based on Russell's circumplex model [10], the variable "nervous" seems to have a negative correlation with productivity. The last variable "tired" also shows a significant positive correlation with productivity.

The regression model, that best explains productivity, is shown in Table 6.2.

The regression model has an R-squared of 0.079 and an adjusted R-squared of 0.067 (p-value is 2.772 E–08). Considering that orders assigned to the employees are not always homogeneous, for example, they may differ regarding the difficulty degree, the personal factors influencing individual productivity and other possible confounding effects, a result of less than 10% is acceptable.

Findings show that the only positive coefficient is "team leader connection", while the other coefficients negatively affect productivity. Performance seems to improve with the increase of connections between the operators and the team leader and if people work in an environment that allows them to be less nervous.

Table 6.1 Correlations with productivity

Variables	Correlations	p-value
Worked_hours	−0.143	3.00 E–04
Team_leader_connection	0.179	6.50 E–05
Nervous	−0.123	5.00 E–04
Tired	0.091	4.90 E–02

Table 6.2 Regression model with productivity as dependent variables

Variables	Coefficient	p-value	Statistical index	Value
Worked_hours	−0.0578	3.93 E–03	Multiple R-squared	0.0787
Team_leader_connection	0.1169	2.95 E–07	Adjusted R-squared	0.0667
Stressed	−0.1079	1.92 E–03	Overall p-value	2.77 E–08
Nervous	−0.2504	3.39 E–03		
Week_day	−0.0381	0.333		
Team_leader_presence	0.1095	0.2843		
Luminosity	−0.088	0.0798		
Hours	8.31 E–04	0.9325		

We controlled for the "week day" because the day of the week can characterize the type and complexity of work and intensity of operations; "team leader presence" because it helps to verify that the variable "team leader connection" is not connected to the physical availability (presence or absence) of the team leader in the warehouse; "luminosity" because usually lighting can contribute to improving workplace condition and overall operation performance; "hour" to double check the "worked_hours" variable. We note that "worked_hours" has a greater coefficient than "hour"; therefore it can be assumed that it is intended as a measure of the tiredness of an operator.

Discussion and Conclusions

This work explores the human and environmental factors affecting the productivity of employees in a semi-automated warehouse by using wearable sensors. As an addition, it methodologically contributes to empirically testing the innovative hardware and software for data collection in this specific setting.

Preliminary results show that, as expected, individual productivity depends on the worked hours, and hence is affected by the fatigue of operators, but it also seems to be influenced by coordination mechanisms and by individual variables such as nervousness and stress of employees. Specifically, an increase in the connection of operators with the team leader, which seems to act as a facilitator of operations, increases the productivity. Also, a high level of stress and nervousness leads to a decrease in individual productivity. Finally, no clear evidence about the role of environmental factors emerges in the experiment.

Some managerial implications can be derived, such as the adoption of part-time workers to reduce the amount of worked hours and avoid high fatigue of employees, as well as incentivizing effective collaboration and interactions of operators with the team leader. While it is difficult to confirm a direct causal relationship between human and environmental factors and operators' productivity, we emphasize the opportunity for managers to care for operators' needs and working environmental conditions in order to improve performance.

Further works should improve this study by extending the time window for data collection in order to cover more variability in the handling activities, by increasing the number of observed workers and by extending the investigation to different warehouses with different type of goods, automation level, and so on. Also, a greater number of environmental sensors could be adopted to cover the areas of a warehouse to provide more detailed insight on the role of environmental factors.

Acknowledgements The authors want to thank the workers of the logistics hub that agreed to be part of the study. They have been informed both about the data collection and the experiment; moreover, they formally accepted to be part of the study through signing a related document.

The collected data have been used in accordance to the national standard (art. 13, D.Lgs 196/2003 and its amendments, Italy) regarding the privacy of personal data.

References

1. M. Napolitano, 2012 warehouse/DC operations survey: mixed signals. Mod. Mater. Handl. **51**(11), 48–56 (2012)
2. E. Grosse, C. Glock, M. Jaber, W. Neumann, Incorporating human factors in order picking planning models: framework and research opportunities. Int. J. Prod. Res. **53**(3), 695–717 (2015)
3. E. Grosse, H. Glock, P. Neumann, Human factors in order picking: a content analysis of the literature. Int. J. Prod. Res. **55**(5), 1260–1276 (2017)
4. P.A. Gloor, A. Fronzetti Colladon, F. Grippa, P. Budner, J. Eirich, Aristotle said "Happiness is a state of activity"—Predicting mood through body sensing with smartwatches. J. Syst. Sci. Syst. Eng. **27**(5) (2018)
5. Z.S. Maman, M.A.A. Yazdi, L.A. Cavuoto, F.M. Megahed, A data-driven approach to modeling physical fatigue in the workplace using wearable sensors. Appl. Ergon. 515–529 (2017)
6. W.P. Neumann, J. Dul, Human factors: spanning the gap between OM and HRM. Int. J. Oper. Prod. Manag. **30**(9), 923–950 (2010)
7. D. Aloini, R. Dulmin, V. Mininno, A hybrid fuzzy-PROMETHEE method for logistics service selection: design of a decision support tool. Int. J. Uncertain. Fuzziness Knowl. Based Syst. **18**(04), 345–369 (2010)
8. D. Aloini, R. Dulmin, V. Mininno, S. Ponticelli, Key antecedents and practices for Supply Chain Management adoption in project contexts. Int. J. Project Manage. **33**(6), 1301–1316 (2015)
9. P. Leather, D. Beale, L. Sullivan, Noise, psychosocial stress and their interaction in the workplace. J. Environ. Psychol. **23**(2), 213–222 (2003)
10. J.A. Russell, A circumplex model of affect. J. Pers. Soc. Psychol. **39**(6), 1161–1178 (1980)

Part II
Emotions and Morality

Chapter 7
Heart Beats Brain: Measuring Moral Beliefs Through E-mail Analysis

Peter A. Gloor and Andrea Fronzetti Colladon

I believe that every human mind feels pleasure in doing good to another.
Thomas Jefferson

Abstract Moral beliefs are at the heart of governing a person's behavior. In this paper, we introduce a way to automatically measure a person's moral values through hidden "honest" signals in the person's e-mail communication. We measured the e-mail behavior of 26 users through their e-mail interaction, calculating their seven "honest signals of collaboration" (strong leadership, balanced contribution, rotating leadership, responsiveness, honest sentiment, shared context and social capital). These honest signals—in other words, how they answered their e-mails—explained 70% of their moral values measured with the moral foundations survey. In particular, the more positive and less emotional they were in their language, the more they cared about others. We verified the results with a larger e-mail dataset of 655 employees of a services firm, where structural and temporal honest signals explained 67% of emotionality.

Introduction

In this paper, we illustrate the link between moral values and emotional behavior predicted through e-mail. In particular, we show that communication patterns measured through e-mail interaction correspond with the moral values of a person.

P. A. Gloor (✉)
MIT Center for Collective Intelligence, Cambridge, MA, USA
e-mail: pgloor@mit.edu

A. F. Colladon
University of Perugia, Perugia, Italy
e-mail: andrea.fronzetticolladon@unipg.it

© Springer Nature Switzerland AG 2020
A. Przegalinska et al. (eds.), *Digital Transformation of Collaboration*,
Springer Proceedings in Complexity,
https://doi.org/10.1007/978-3-030-48993-9_7

Emotions Control Moral Values

Former US Vice President Joe Biden ran into difficulties by becoming too emotional and touchy-feely with his supporters, while Senator Elizabeth Warren got a lot of criticism for claiming native American ancestry. In these instances, the politicians followed their feelings over rational behavior. Frequently people are not aware of their emotions. While we think that we are rational creatures who will make decisions based on reason, the opposite is true. People will make emotional decisions, and then find rational reasons to justify their emotional judgments [1]; this means that a posteriori reasoning is applied to justify a priori emotional decisions.

There is a bidirectional link between emotions and morals. Morals give an ethical compass to individuals guiding them in their decisions, to decide what is right or wrong. While moral behavior is commonly assumed to be a rational process, in reality it is driven by emotions. Specifically, moral emotions influence the link between moral standards and behavior [2, 3]. Past research in the cognitive and neurobiological sciences suggested that emotions are necessary, sometime sufficient, for moral judgment [4, 5, 6, 7]. This means that acting on moral beliefs is controlled through our emotions [8]. Emotions with negative valence such as shame, guilt, embarrassment and disgust are key drivers for what we find morally acceptable or not. Also, on the positive side, emotions such as gratitude, pride and moral elevation, inspiring others to act virtuously, are the trigger that makes us feel good, leading to rational justification of morally positive behavior. There is a strong link between moral standards and moral behavior. Indeed, "*as the self reflects upon the self, moral self-conscious emotions provide immediate punishment (or reinforcement) of behavior [...] When we sin, transgress, or err, aversive feelings of shame, guilt, or embarrassment are likely to ensue. When we "do the right thing," positive feelings of pride and self-approval are likely to result*" [3]. Similarly, consumer behavior can be triggered by moral emotions, as a response to company actions [9]. Already Thomas Jefferson assumed in the late eighteenth century that witnessing acts of charity and benevolence by others would instigate a yearning by individuals to behave in a similar positive way.

Nurturing Positive Emotions Makes Us Happy

Positive emotions enhance psychological functioning [10], increase life satisfaction and make us happy [11]. According to the bestselling book "Aging Well" [12], there are two points that get us to old age: attitude and gratitude. In more detail, Vaillant identifies five key factors for happy aging: (1) maintaining stable positive relationships, (2) good coping skills in adversity, (3) keeping a healthy weight and exercising regularly, (4) not smoking and only drinking alcohol in moderation, and (5) pursuing continuing education. These five life-changing and life-extending factors require individual resolutions, which are triggered by emotional decisions. Whether it is accepting the first cigarette or bottle of beer at a fraternity party, or the decision to propose marriage to a loved one, or to stop smoking or drinking alcohol, emotions decide whether to cave in, or resist the short-term temptation for long-term gratification and what Aristotle calls in his Nicomachean Ethics "higher happiness". In this sense, individual resolution is important as the duration of people's positive feelings impacts wellbeing more than the intensity of these feelings [13].

Trusted Peers (Re)Define Our Belief System

We do not make decisions in isolation, but influenced by others. If people trust somebody, they will follow their advice. In medieval villages, people did what their trusted person of authority, be it priest or village elder, told them. Today, they trust their friends, sometimes even online friends on Facebook and on other social media [14]. We know by work on social capital that primary relationships, that is, strong ties, are key enablers of trust [15, 16]. Social networking sites may augment and reinforce pre-existing strong ties, based on personal face-to-face encounters, thus contributing to the shaping of emotions and moral beliefs.

However, people also make decisions based on their belief system, which, as we have just seen above, is based on their emotions. For instance, the decision to trust a stranger is also an emotional decision based on intuition. In our daily life, we are constantly presented with new claims asking for intuitive decisions driven through emotions. Be it the touchy-feeliness of Joe Biden, or the native American heritage of Elizabeth Warren, one has to decide to either accept a new claim as truth, or reject it as a lie. If the new claim is introduced by somebody we trust, we usually accept it as truth. Figure 7.1 simplifies the process taking place when deciding if a claim is interpreted as truth or lie.

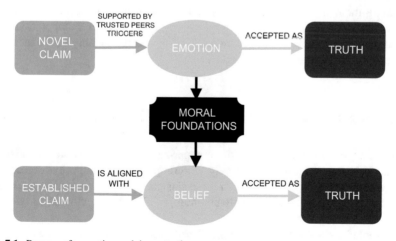

Fig. 7.1 Process of accepting a claim as truth

It is therefore reasonable to assume that what somebody tells us will influence our emotions, and thus our decisions [17], which are steered by our moral belief system. Pentland [18] defines "honest signals" as personal patterns that an individual demonstrates while completing a task without being consciously aware of it. In this project, we show that the honest signals in e-mail, calculated through semantic and social network analysis, predict people's moral values.

Methods

In a first case study, we analyzed the mailbox of a co-organizer of a scientific event, studying his e-mail interaction with 26 participants of the event. To track their interactions, we calculated the "seven honest signals of collaboration" [19] for each of the participants from his e-mail archive. These honest signals (strong leadership, balanced contribution, rotating leadership, responsiveness, honest sentiment, shared context and social capital) have been shown in earlier works [20, 21] to be predictive of the several dependent variables such as customer satisfaction or work engagement.

The 26 participants of the event also took the Moral Foundations survey [22]. It measures the moral values of the respondent in five categories (care, fairness, loyalty, authority and sanctity). For the analysis, these five foundations can be grouped into two higher-order clusters: care and fairness, and loyalty, authority and sanctity. In addition, participants took the Schwartz values test [23], which measures moral attitudes in the two aggregated dimensions conservation and transcendence. Conservation includes the values of security, conformity and tradition. Transcendence is composed of benevolence and universalism.

In an additional study, we compared the honest signals of 655 employees of a firm calculated through their e-mail, to show the link between emotions and temporal and structural e-mail communication patterns. We analyzed two months of e-mail, including the meta information such as sender, recipients, timestamp and subject line of the messages, to compute the seven honest signals of collaboration [19]. Our dependent variable in this second analysis is the emotionality of the messages calculated from the subject line.

Results: E-mail Behavior Reflects Moral Foundations

Table 7.1 shows the results of comparing both the moral foundations values harm/care, fairness/reciprocity, in-group loyalty, authority/respect and purity/sanctity, as well as the Schwartz value clusters conservation and transcendence with the seven honest signals of collaboration [19]. We find significant correlations for almost all moral values.

As Table 7.1 shows, there is a strong link between the number of messages sent and individuals' harm/care score. We also find a positive association of rotating leadership

Table 7.1 Pearson's correlation coefficients of honest signals and individual differences

	1	2	3	4	5	6	7	8	9	10	11	12	13	14	15	16	17	18	19	20	21
1 Harm Care	1.000																				
2 Fairness Reciprocity	.615**	1.000																			
3 In-Group Loyalty	0.155	-0.069	1.000																		
4 Authority Respect	-0.289	-.416*	.393*	1.000																	
5 Purity Sanctity	0.095	-0.313	.394*	.641**	1.000																
6 Conservation	-0.173	-.446*	0.174	.624**	.590**	1.000															
7 Trascendence	0.303	0.318	-0.154	-0.129	0.113	0.196	1.000														
8 Sentiment	0.259	0.357	0.077	-0.056	0.023	0.016	0.149	1.000													
9 Alter ART	-0.125	0.134	-0.209	-0.040	0.002	-0.077	0.335	-0.270	1.000												
10 Ego ART	0.172	-0.254	-0.141	0.100	0.086	.405*	0.163	0.014	-0.084	1.000											
11 Alter Nudges	0.056	-0.286	0.423	.479*	0.184	0.297	0.063	-0.059	-0.057	0.155	1.000										
12 Ego Nudges	-0.146	-0.184	0.092	0.321	0.351	0.257	-0.053	0.116	-0.347	0.030	-0.263	1.000									
13 Messages sent	.404*	0.296	0.241	-0.197	-0.059	-0.245	0.232	-0.121	-0.020	-0.101	0.138	-0.130	1.000								
14 Messages received	0.381	0.244	0.198	-0.202	0.018	-0.192	0.258	0.068	-0.029	-0.097	0.046	-0.059	.916**	1.000							
15 Contribution index	0.282	0.242	0.098	-0.084	-0.060	-0.034	0.041	.363*	0.149	-0.312	0.039	-0.337	0.101	-0.102	1.000						
16 total influence	0.386	0.295	0.175	-0.222	-0.111	-0.233	0.261	-0.103	-0.010	-0.090	0.133	-0.134	.987**	.935**	0.070	1.000					
17 Betweenness centrality oscillation	.394*	0.322	0.222	-0.143	-0.084	-0.189	0.210	-0.120	0.185	-0.004	0.246	-0.232	.769**	.482**	.399*	.718**	1.000				
18 Betweenness centrality	0.378	0.227	0.177	-0.218	0.019	-0.189	0.268	-0.029	-0.036	-0.091	0.030	-0.055	.870**	.993**	-0.135	.901**	.394*	1.000			
19 Degree centrality	.398*	0.253	0.218	-0.194	0.019	-0.188	0.268	-0.028	-0.021	-0.078	0.083	-0.084	.933**	.992**	-0.061	.953**	.538**	.984**	1.000		
20 Complexity	0.297	.410*	-0.025	-0.197	-0.181	-0.025	0.033	.550**	-0.068	-0.093	-.543**	0.147	0.242	0.128	.717**	0.221	.379*	0.098	0.161	1.000	
21 Emotionality	-0.348	-0.197	-0.147	-0.041	0.034	-0.206	-0.042	-.456*	0.422*	-.431*	-.646**	0.061	-0.194	-0.089	-.664**	-0.170	-0.306	-0.060	-0.105	-.768**	1.000
22 Reach2	0.098	-0.070	0.137	-0.014	0.204	.384*	0.178	0.324	0.018	-0.082	0.050	-0.117	0.324	0.256	.475**	0.310	0.356	0.228	0.281	.634**	-.631**

* p < .05; ** p < .01; *** p < .001

(betweenness oscillations) and of degree centrality with this score reflecting virtues of kindness, gentleness and nurturance. Those who have the ability of being more caring toward others, send more messages, have more direct social contacts and rotate in the network without maintaining static positions. The fairness/reciprocity score, on the other hand, is positively associated with language complexity. People who try to be fair use more complex language. People who score high on the authority/respect scale receive more nudges by their peers. This means that individuals who value authority and respect need to get more nudges from their peers until they respond.

With regards to the dimension of conservation of the Schwartz test, we find that those who care more about security, conformity and tradition answer e-mails faster and have higher social capital as captured in the variable Reach2 [19]. Reach2 defines the number of people that one can reach in two degrees of separation. In other words, this variable does not directly measure how many friends somebody has, but how many friends her friends have. It has been shown to be a good predictor of social and financial capital [24].

We additionally find that the honest signals of communication can predict the moral values of a person. As the regression models for the Schwartz values are somewhat less accurate, we present the regressions for the moral values. The regression models with the best fit are shown in Table 7.2, illustrating which variables matter the most while predicting each individual trait.

We find, for instance, that the more positive and the less emotional people are, the more they care about others. A similar behavior (positive and non-emotional language) is also indicative of people who value sanctity and purity. On the other hand, the more people nudge others and are nudged themselves by e-mail, they more they value authority and respect.

Table 7.2 Moral foundations test—regression models

Predictors	Dependent variable				
	Harm/Care	Fairness reciprocity	In-group loyalty	Authority respect	Purity sanctity
Sentiment	74.3173**	65.5398**			62.8351**
Alter ART	0.2245**	0.1179*		0.0976^	0.2483**
Ego ART				0.1260^	0.2049**
Alter nudges	−18.2495**	−6.9631^		10.7797*	
Ego nudges	8.7743**			8.2856*	16.2718**
Messages sent					0.0173**
Messages received		0.0126**	0.0082^	0.0096^	
Contribution index					
total influence			−0.0660^		−0.1453**
Betweenness centrality oscillation	0.0992***			-0.1109*	
Betweenness centrality		−0.0004**	−0.0004*	-0.0006*	
Degree centrality			0.1930*	0.2010*	
Complexity					
Emotionality	−176.6017***		−35.3722^		−52.3369*
Reach2			−0.2093^		
Constant	18.3043	−17.9746	157.2252*	−16.7616^	−49.0391**
Adjusted R-squared	0.6246	0.4514	0.3812	0.5512	0.6396

$^\wedge p < 0.1$; $^* p < 0.05$; $^{**} p < 0.01$; $^{***} p < 0.001$

Verifying the Link Between Emotion and E-mail Dynamics and Structure

To further demonstrate the link between emotions and e-mail behavior, we analyzed an e-mail archive with two months' worth of e-mail of 655 employees of a professional services firm, where we compared the structural and dynamic honest signals with their emotionality calculated from subject lines. This means that different from the first analysis, this time we only had the words contained in the subject line. However, in previous work the predictive power of this approach had been illustrated [21], as sentiment, emotionality and complexity of subject line and email body are correlated if the sample is big enough.

Table 7.3 Predicting emotionality—regression model

Variable	Coefficient
Betweenness centrality oscillation	-0.0026^{***}
Degree centrality	0.0002^{***}
Contribution index	-0.0604^{***}
Reach2	-0.00002^{***}
Constant	0.4138^{***}
Adjusted R-squared	0.6681

$^{***}p < 0.001$

We find that 67% of the emotionality of an employee is explained by rotating leadership defined as betweenness centrality oscillation, central leadership defined as degree centrality, contribution index, and social capital defined as reach-2 (see Table 7.3) [19]. In other words, there is a hidden strong link between the structure and dynamics of communication, and the contents of the communication. This goes back to the original definition of "Honest Signals" [18] which give always what somebody really thinks without explicitly saying it. Here we have shown that this is even true for e-mail communication: how central somebody is in the network, how much she sends compared to receiving e-mail, and how many times she changes her network position, and how popular her friends are, gives away her emotional state.

Discussion and Conclusions

Getting e-mails with an established claim from a trusted source will make the recipients interpret it positively, eliciting a different type of response—based on their moral foundations—than if they do not trust the source. The same is true if one is getting e-mails with novel claims. They will trigger different types of emotional responses based on if the recipients trust the source, and on their moral foundations. Either way, in this research we have shown that analyzing individuals' e-mails can reveal their moral foundations. We have measured the seven honest signals of communication to characterize the e-mail behavior of different people. At the same time, we asked the people whose e-mail communication was analyzed to take the Schwartz value test [25] and the moral foundations test [22], finding a significant link between e-mail behavior and moral values. There is a strong link between ethics and emotions, both on the business and the psychological side [26, 27, 3]. The ethical and moral values of a person are given away by their "honest signals" as expressed through their e-mail behavior. This e-mail behavior is predictive of emotions, which are predictive of the ethical and moral values of a person.

While the results we found are intriguing, this is preliminary work motivating much further research. Our study should be repeated in a broader setting with more participants. Additionally, in the second study of this research, we take subject lines as a proxy of email bodies—consistent with past research which showed that people sentiment and emotionality measured on subject lines are correlated with the same metrics calculated on the email body [28]. Nevertheless, this study should be repeated also accessing e-mail bodies and not only subject lines, for a more accurate assessment of honest signals related to language use.

We have shown that how someone communication in e-mail can predict their moral values and emotionality. These insights can be applied to virtual mirroring [20, 21], providing an automated way for making the moral values of individuals more obvious to them, thus assisting them for better self-management and self-understanding, ultimately leading to higher happiness in the Aristotelian sense.

All procedures performed in studies involving human participants were in accordance with the ethical standards of the institutional and/or national research committee with the Helsinki declaration and its later amendments or comparable with ethical standards.

References

1. D. Ariely, *Predictably Irrational* (Harper Audio, 2008)
2. B. Huebner, S. Dwyer, M. Hauser, The role of emotion in moral psychology. Trends Cogn. Sci. **13**(1), 1–6 (2009). https://doi.org/10.1016/j.tics.2008.09.006
3. J.P. Tangney, J. Stuewig, D.J. Mashek, Moral emotions and moral behavior. Annu. Rev. Psychol. **58**, 345–372 (2007)
4. J.D. Greene, An fMRI investigation of emotional engagement in moral judgment. Science **293**(5537), 2105–2108 (2001). https://doi.org/10.1126/science.1062872
5. J.D. Greene, L.E. Nystrom, A.D. Engell, J.M. Darley, J.D. Cohen, The neural bases of cognitive conflict and control in moral judgment. Neuron **44**(2), 389–400 (2004). https://doi.org/10.1016/j.neuron.2004.09.027
6. J. Prinz, The emotional basis of moral judgments. Philos. Explor. **9**(1), 29–43 (2006). https://doi.org/10.1080/13869790500492466
7. S. Schnall, J. Haidt, G.L. Clore, A.H. Jordan, Disgust as embodied moral judgment. Personal. Soc. Psychol. Bull. **34**(8), 1096–1109 (2008). https://doi.org/10.1177/0146167208317771
8. J. Haidt, in *The Righteous Mind: Why Good People are Divided by Politics and Religion* (Vintage, 2012)
9. S. Grappi, S. Romani, R.P. Bagozzi, Consumer response to corporate irresponsible behavior: moral emotions and virtues. J. Bus. Res. **66**(10), 1814–1821 (2013). https://doi.org/10.1016/j.jbusres.2013.02.002
10. I.B. Mauss, A.J. Shallcross, A.S. Troy, O.P. John, E. Ferrer, F.H. Wilhelm, J.J. Gross, Don't hide your happiness! Positive emotion dissociation, social connectedness, and psychological functioning. J. Personal. Soc. Psychol. **100**(4), 738–748 (2011). https://doi.org/10.1037/a0022410
11. M.A. Cohn, B.L. Fredrickson, S.L. Brown, J.A. Mikels, A.M. Conway, Happiness unpacked: positive emotions increase life satisfaction by building resilience. Emotion **9**(3), 361–368 (2009). https://doi.org/10.1037/a0015952
12. G.E. Vaillant, *Aging Well: Surprising Guideposts to a Happier Life from the Landmark Study of Adult Development*. Little, Brown. Saunders DS (1976) The biological clock of insects. Sci. Am. **234**(2), 114–121 (2008)
13. E. Diener, E. Sandvik, W. Pavot, Happiness is the frequency, not the intensity, of positive versus negative affect, in *Assessing Well-Being*, ed. by E. Diener (2009), pp. 213–231. https://doi.org/10.1007/978-90-481-2354-4_10

14. A. Majchrzak, in *Unleashing the Crowd: Collaborative Solutions to Wicked Business and Societal Problems* (Springer Nature, 2019)
15. M.J. Granovetter, The Strength of Weak Ties. Am. J. Sociol. **78**(6), 1360–1380 (1973). https://doi.org/10.1086/225469
16. D. Krackhardt, The strength of strong ties: the importance of philos in organizations, in *Networks and Organizations: Structure, Form, and Action*, ed. by N. Nohria, R.G. Eccles (Harvard Business School Press, Boston, MA, 1992), pp. 216–239
17. P. Watzlawick, J.H. Weakland, R. Fisch, *Change: Principles of Problem Formation and Problem Resolution* (Norton & Company, New York, NY, 2011)
18. A. Pentland, *Honest Signals: How They Shape Our World* (MIT Press, 2010)
19. P.A. Gloor, *Sociometrics and Human Relationships: Analyzing Social Networks to Manage Brands, Predict Trends, and Improve Organizational Performance* (Emerald Publishing Limited, 2017)
20. P. Gloor, A.F. Colladon, G. Giacomelli, T. Saran, F. Grippa, The impact of virtual mirroring on customer satisfaction. J. Bus. Res. **75**, 67–76 (2017)
21. P. Gloor, A.F. Colladon, F. Grippa, G. Giacomelli, Forecasting managerial turnover through e-mail based social network analysis. Comput. Hum. Behav. **71**, 343–352 (2017)
22. J. Graham, J. Haidt, S. Koleva, M. Motyl, R. Iyer, S.P. Wojcik, P.H. Ditto, Moral foundations theory: the pragmatic validity of moral pluralism, in *Advances in Experimental Social Psychology*, vol. 47 (Academic Press, 2013), pp. 55–130
23. S.H. Schwartz, An overview of the Schwartz theory of basic values, in *Online Readings in Psychology and Culture* (2012)
24. B. Hadley, P.A. Gloor, S.L. Woerner, Y. Zhou, Analyzing VC influence on startup success: a people-centric network theory approach, in *Collaborative Innovation Networks* (Springer, Cham, 2018), pp. 3–14
25. M. Lindeman, M. Verkasalo, Measuring values with the short Schwartz's value survey. J. Personal. Assess. **85**(2), 170–178 (2005)
26. Y. Lurie, Humanizing business through emotions: on the role of emotions in ethics. J. Bus. Ethics **49**(1), 1–11 (2004)
27. E. Pulcini, What emotions motivate care? Emotion Rev. **9**(1), 64–71 (2017)
28. A. Fronzetti Colladon, P.A. Gloor, Measuring the impact of spammers on e-mail and Twitter networks. Int. J. Inf. Manag. **48**, 254–262 (2019). https://doi.org/10.1016/j.ijinfomgt.2018.09.009

Chapter 8
Identifying Virtual Tribes by Their Language in Enterprise Email Archives

Lee Morgan and Peter A. Gloor

Abstract The rise of online social networks has created novel opportunities to analyze people by their hidden "honest" traits. In this paper we suggest automatic grouping of employees into virtual tribes based on their language and values. Tribes are groups of people homogenous within themselves and heterogenous to other groups. In this project we identify members of digital virtual tribes by the words they use in their everyday language, characterizing email users by applying four macro-categories based on their belief systems (alternative realities, personality, recreation, and ideology) developed in earlier research. Each macro-category is divided into four orthogonal categories, for instance "Alternative Realities" includes the categories "Fatherlanders", "Treehuggers", "Nerds", and "Spiritualists". We use the *Tribefinder* tool to analyze two email archives, the individual mailbox of an active academic and corporate consultant, and the Enron email archive. We found tribes for each user and analyzed the communication habits of each tribe, showing that members of different tribes significantly differ in how they communicate by email. This demonstrates the applicability of our approach to distinguish members of different virtual tribes by either language used or email communication structure and dynamics.

Introduction

In today's age of alternative realities, different groups in society look at the same underlying evidence as either fact or fiction. In this paper we apply a system we developed earlier to find these groups—virtual tribes—in the corporate world. Our goal is to identify virtual tribes among employees of a company to better understand the different value systems motivating the members of the organization.

Tribes are groups of people that share common ideas, thoughts, and emotions [1]. In other words, they are people who have strong cultural, emotional, and ideological links to each other, creating a sense of community [1]. There are many types of tribes,

L. Morgan · P. A. Gloor (✉)
MIT Center for Collective Intelligence, 245 First Street, Cambridge, MA 02142, USA
e-mail: pgloor@mit.edu

© Springer Nature Switzerland AG 2020
A. Przegalinska et al. (eds.), *Digital Transformation of Collaboration*,
Springer Proceedings in Complexity,
https://doi.org/10.1007/978-3-030-48993-9_8

and they can vary in size and role [2]. However, most of the literature surrounding tribes has seen their use in marketing [3]. Consumer tribes have emerged as an important part of a firm's success. There is limited use of tribes being used for managing human resources, as we will be proposing in this paper.

Human resource management has been taking strides toward the use of analytics for better understanding the wants and needs of employees. In the past years, human resource research based on mining data has become a notable field, with dozens of studies emerging [4]. This has permeated into industry as well, with multiple firms employing these techniques [5], for instance using email for data-driven human resources management [5].

Until now, the concept of virtual digital tribes has yet to be used. In earlier work, strategic benefits from using the concept have been shown, like increasing the happiness of workers using virtual mirroring and identifying different emotions through tribes [3]. Until now it has been difficult to automatically identify tribes instantly, without using surveys or other manual tools, but based on systematically identifying tribes based on their activity online. As a result, it has been difficult to identify and classify membership in tribes on a large scale. Owing to the rise of data-driven human resource management, social media, and email, a different, digitally-based, tribe identification method has been developed. Online communities can be easily formed based on a common idea or interest, and can have the same positive and negative implications as tribes that are not based on the internet, though over the internet they might have a greater impact due to the ease of access and spreading of information [6]. We call these new tribes virtual, electronic, or e-tribes [7].

Tribefinder is a novel system that uses artificial intelligence and machine learning to identify the tribes of users based on social media data [7]. While originally created to be used with data from Twitter, it can also operate on other forms of media, like email [7]. Tribefinder works through the use of word embeddings and long short-term memory (LSTM) [8]. It currently determines tribes using the words in their messages. More specifically, it finds the different types of tribes and their leaders on Wikipedia, then looks at the language of the leaders on Twitter [7]. People are assigned to tribes if their word usage is like that of the aggregate of all "leaders" of a tribe [7]. As a proof of concept for applying Tribefinder for email, this paper uses Tribefinder to determine the tribes in a personal inbox and the Enron dataset [9]. Tribefinder works with multiple macro-categories of tribes, with users fitting into a specific tribe under each macro-category. This paper will work with the *Alternative Realities, Personality, Recreation,* and *Ideology* macro-categories [7]. Each tribe has specific traits, and this paper will look at and compare them between the different tribes. The traits, or honest signals, that this paper uses are related to productivity, connectivity, complexity, and communication habits of each tribe [10].

This paper advances the current research as it applies *Tribefinder* to human resources (HR) management and email. It will also show the differences in the traits of each email tribe, in order to tease out the characteristics of each tribe. This adds onto data-driven management, as it offers a novel way to analyze the data of employees and boost their productivity. This can be done through *virtual mirroring,* which mirrors back the communication habits of persons, causing internal reflection [10]. As a result, there can be an increase in customer satisfaction and overall productivity.

Theoretical Background

Virtual Tribes

Tribes are effectual groups that are not held together by formal societal constructs, but instead, a common emotion, belief, or ideology glues their members with each other [11]. These tribes can be heterogeneous, meaning that members have differences in their ages, incomes, gender, races, and social status; the most important factor about deciding a tribal affiliation is a common belief [1]. The postmodern society contains a large amount of these invisible micro-groups, which all share strong emotional links [2]. Moreover, "tribe", as a word, hints at seemingly ancient and archaic values, like "a local sense of identification, religiosity, syncretism, group narcissism etc." ([11], p. 67).

Humans typically choose which tribes to associate themselves with through their actions and behaviors [12]. This is called the self-categorization theory; it occurs due to one's access and fit to a tribe [13]. Fit is the extent to which tribes reflect realistic societal groups and statuses. A high fit would indicate minimal intra category differences and minimize inter category similarities [14]. Access is simply the ease of joining a tribe and its proximity to an individual. If one has more access to a tribe, they are more likely to categorize with it [14]. It should be noted that self-categorization theory states that the process of finding a tribe can change based on the situation and is always based on the perspective of the perceiver [14]. Other factors that would influence one's social categorization would be benefits to one's identity, place in society, a stronger sense of community, emotional links, and ethnic partiality [15–17]. Additionally, people can identify with the members of one or more tribes [18]. This is because humans need to express separate parts of their identities, and one tribe alone cannot typically do this; humans need multiple tribes to accommodate for different aspects of their identities [18]. An example of the behaviors of tribes is an "anchoring event", where tribe members meet in public areas and perform ritual acts [19]. These anchoring events are essential for tribes to have consistent and sustained membership as they enforce the key ideals and values of tribes [19]. However, it should be noted that there is a spectrum when it comes to engagement in "anchoring events" [1]. On one side, there are sympathizers, who have a limited amount of interest in the tribe, and on the other side, there are practitioners, whose identities are based on the tribe and who engage with it daily [1]. For these reasons, tribalism is emerging in our society, and today's tribes are highlighted by an important duality: the tribe influences its members, and at the same time, the members define their tribe [20, 17]. Moreover, traditional tribes have also shifted to virtual tribes or e-tribes [21, 22]. This is due to the rise of social media and the internet as a whole: there are many new forums and sites for tribes to be fostered and created [6]. Tribes have been found and researched on sites like Twitter, due to the volume and availability of public messages on the site [7].

Tribes in Human Resources Management

Data-driven human resource management can be defined as the use and mining of data coming from employees and customers of a firm, and implementing models and solutions based on the data in the company [23]. Indeed, many companies are now utilizing data instead of a manager's "gut instinct" when it comes to making decisions [24]. This is accompanied by a data revolution, where companies can now easily collect and aggregate data [24]. This practice has brought in substantial benefits as the decision-making that comes from mining data can cause a 5–6% increase in firm productivity [24].

Despite the rise of data-driven HR (human resources), tribes and community organization are rarely utilized when it comes to HR management [5, 25]. They have, however, seen their use in the marketing world [7]. Tribes for specific firms, or *brand communities* have been used to easily and quickly spread information about products to consumers [7]. Because of this, the formation of consumer tribes has been identified as critical to the survival of firms at any stage of their development [12].

Though emails may seem less like a social network and more of a communication tool, research based on mining data from email databases proves the opposite [26]. Emails have also been used as a source to mine data [26]. In professional environments, emails represent a typical social network and exhibit "long-tailed, small-world" traits [26]. There are varying levels of participation and leadership within this space as well: a few members send the vast majority of the emails, and there seems to be a hierarchy in the social network [26]. Thus, email provides a useful substrate for discovering the tribal affiliations of individuals and groups. For example, a study looked at a firm's emotional tribes and found conclusive results which could be virtually mirrored back to the employees to increase the productivity and happiness of the workers [3, 10].

Methods

Challenges in Finding Tribes

Tribes can be compared with the elementary particles of quantum physics as they are difficult to pinpoint due to their fuzzy and ever-changing nature [19]. Thus, even though there have been many methods of identifying tribes, like interviews, focus groups, surveys, ethnographic, and netnographic approaches, there has been no way to rapidly and automatically identify tribes based on their traits [7]. These manual methods provide a deep understanding of tribal characteristics. In the past decades, tribal studies have used limited surveying methods, like making surfers fill out questionnaires and studying small groups of adult record collectors [18, 27]. For analyzing e-tribes with millions of members, these methods are impractical and cumbersome [7]. Rather, for a large company that wanted to find tribal attributes of its employees, it would be easier to analyze email databases.

Tribefinder

In order to solve these issues around manual identification of tribes, *Tribefinder* discovers tribes based on text data [7]. *Tribefinder*, which until now has been mainly been used on Twitter, categorizes people into one tribe for each specific macro-category; since there are multiple macro-categories, people can belong to multiple tribes [7]. *Tribefinder's* analysis of tweets and emails extracts data on multiple ideas and leaders. It outputs tribal affiliations for each specific user allowing researchers or managers to find typically unnoticeable traits that distinguish individuals. It should be noted that the macro-categories that *Tribefinder* can output are not rigid. In previous instances, its outputs were macro-categories like *Alternative Realities*, *Ideologies*, and *Personalities*, but the user of *Tribefinder* can create different tribes based on their own specifications [7]. For instance, *Tribefinder* has been used to create a "Bernie Sanders Tribe" and a "Donald Trump Tribe" and sort Twitter users into them [10].

Tribefinder includes two functions: tribe allocation and tribe creation. With the tribe creation function the user can create macro-categories and specific tribes within them. Tribe allocation assigns tribal affiliations to people based on their characteristics. This paper uses the tribe allocation process as the macro-categories are already determined.

In order to create a new tribal macro-category, the user first has to find a group of key individuals that are representative of each tribe within the tribal macro-category, the "tribe leaders" [7]. For instance, the "Bernie Sanders Tribe" could have Bernie Sanders and some of his most ardent supporters and campaign managers as its leaders. After this, *Tribefinder* would find a large sample of individuals similar to the "tribe leaders" based on automatically extracted keywords that would be associated with a certain tribe. After the user would identify key leaders, dozens of similarly self-identified members of the tribe will be proposed as additional tribe leaders. For instance, if the user wanted to find individuals that were part of the "Arts" tribe on Twitter, it would search profiles for biographies, tweets, friends, and followers related to the art in order to find these new tribe members. After this, they are shown to the user, who can choose to include these people as tribe leaders or not.

These results can be demonstrated to the user in two types of charts. The first is a word cloud which shows the most common concepts for a certain tribe and can act

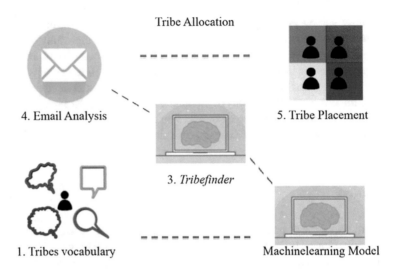

Fig. 1 Tribe allocation diagram

as a suggestion for new keywords. The second is a drawn-out network of members to demonstrate the most connected and well-known members of tribes.

After this, tribe allocation occurs. This begins with TensorFlow deep learning being used to find the key patterns and ideas in the tweets of tribe leaders [28]. This is used to identify textual patterns for each tribe and create a specific set of words for each tribe as well. Then, more deep learning is used to analyze the vocabulary and syntax of tribe leaders in order to be able to connect unconnected individuals to a tribe [28]. Then, using long short-term memory and word embeddings, classifications for specific users can be created. Specifically, one's words in emails or tweets are converted into vectors which are then inputted into the long short-term memory models [29].

Although it has been difficult to run models on sites like Twitter due to a lack of long messages [30], the word embeddings and LSTM used proved to be sufficient (Fig. 1). One limitation is that *Tribefinder* needs a large amount of tweets or emails in order to work. It should be noted that *Tribefinder* does not depend on deep learning models in order to work, as it can use different methods for short text analysis, but it is most accurate when LSTM and word embeddings are used [7].

Tribe Categories

This paper does not focus on developing new tribes, but rather focuses on analyzing the traits of preexisting macro-categories. The macro-categories focused on in this paper are *Alternative Realities, Personalities, Recreations,* and *Ideologies.* These have been created and utilized in previous research on tribes [7].

The *Alternative Realities* macro-category is broken into four groups. The first are fatherlanders, which can be described as extremely patriotic. Their main vision would be a re-creation of the national states from the 1900s. The spiritualism tribe unsurprisingly has a focus on all things spiritual. The nerds are people who believe in seeing advances in technology and strides to the future. The tree huggers are environmentalists and strive to protect nature from phenomena like global warming [7].

The *Personalities* macro-category has four parts as well: stock-traders, politicians, journalists, and risk-takers. Stock-traders have a focus on capital and the economy. "Politicians" are representative of people who use "political language" instead of simply saying the truth. Risk-takers like to make daring decisions (this category has been trained with wingsuit flyers and cave divers), and journalists, other than politicians, use direct language to report actual events.

The *Recreation* macro-category is composed of the fashion, art, travel, and sport tribes. Fashion tribe members focus on the new styles of clothes; the arts tribe has an interest of all types of art like music and painting; the travel tribe enjoys traveling around the world; and the sport tribe enjoys actively engaging in sports [7].

Finally, the *Ideologies* macro-category is made of the liberalism, socialism, capitalism, and complainers tribes. The liberalism tribe focuses on enhancing and protecting the freedom of individuals. The socialism tribe advocates for more government control and intervention in economies. The capitalism tribe is practically the opposite of socialism—it argues for minimal government intervention in markets. The complainers tribe frequently voices their protests to problems they see.

Utilizing Tribefinder: Honest Signals

Tribefinder proved to be powerful as it can be used to discover non-obvious characteristics of employees in a firm. In previous work, it was tested with Twitter with an accuracy rate of 81.2% in the best case and 68.8% in the worst case [7]. It has been used to identify customer's tribal affiliations to see which tribes are more likely to have interest in certain brands, and it has been used in identifying the traits of customer tribes [7]. In this paper, employee tribes will be analyzed through the use of the honest signals [10, 31]. These honest signals identify differences in the activity and language of tribes, which the firms can use to see which tribes are the most positive and active in the workplace.

Honest signals are part of network science. They provide a way of seeing different actors as part of a group instead of observing them as individuals [31]. Honest signals can be seen as seemingly unnoticeable patterns which reveal the goals and key ideals of people to others [31]. They are honest because they are uncontrollable because of being processed unconsciously. These honest signals can be extremely effective as they can predict outcomes in seemingly random situations, like dates and job interviews [31].

The interconnectedness of a tribe is measured through its social network's centrality [32]. There are two measures of centrality used in this paper: betweenness and degree centrality. Degree centrality is simply the amount of people a user sends and receives emails from. Betweenness centrality is the frequency in which a user appears in a path connecting other users. This is computed through finding the shortest paths in a network that connects all network's users to each other, then counting the amount of time one appears in a path connecting two other users [33].

The activity of users was measured through messages sent, contribution index, and rotating leadership. Messages sent is simply the amount of emails sent by an individual. Rotating leadership is the oscillations in betweenness centrality in a specified time period (15 days). This is calculated by finding the number of local maxima and minima in the betweenness curve of an actor [10]. A rotating leader is someone who alternates between being a leader and follower in groups. For example, they would start conversations, then allow the other members of the network to carry them on [34]. The contribution index measures the balance of messages sent and received by a user. It is calculated by subtracting messages received from messages sent and then dividing the result by messages sent added to messages received [10].

The last characteristic analyzed was the language of the tribe members. This was done by finding the average sentiments, emotionality, and language complexity. Average sentiment is the measure of the positivity and negativity of a user's emails. It was calculated using a classifier algorithm and varies between 0 and 1, with 0 being the most negative and 1 the most positive [7]. Average emotionality is the measure of user's deviation from the usual sentiment and is measured as the standard deviation from the mean sentiment [10]. Finally, average complexity measures the complexity of a user's vocabulary. The more varied words one uses, the higher their complexity [10]. All of the honest signals were calculated with Condor [10].

Results

Though *Tribefinder* can be used to create new tribes, this paper works under the framework using the predefined tribes provided by Gloor et al. [7]. These tribes have their notable traits identified through data mining emails and social network analysis. This can be impactful as firms can identify employee tribes that need an increase in sentiment and those that speak with the highest complexity, meaning new ideas are coined. This analysis uses the Enron large dataset [10], which had 1738 users that were placed into tribes. The results from Enron are compared to those of a private email inbox, which had its 20 most active participants placed in tribes. The charts and Table 1 demonstrate the significant differences within tribes and compare the results for the Enron and private emails. The bar charts have error bars of the 95% confidence intervals (Figs. 2, 3, 4, 5).

For the *Alternative Realities* macro-category, the one-way analysis of variance (ANOVA) demonstrates multiple significant differences in honest signals in both email datasets. The Spiritualism tribe has the lowest average complexity in both

Table 1 Differences in honest signals among *Alternative Reality* tribes for email inbox and Enron inbox

Honest signal	Group	Group	Mean difference	*P*-value
Significant comparisons for email data, *N* = 20				
Average complexity	Spiritualism	Nerd	−1.814	0.0331496
		Treehugger	−2.623	0.0018502
Significant comparisons for Enron data, *N* = 1738				
Rotating leadership	Spiritualism	Nerd	−11.67	0.0103851
		Treehugger	−12.47	0.015948
Average sentiment	Nerd	Fatherlander	0.0270	0.0246773
		Treehugger	0.0129	0.0402434
Average complexity	Spiritualism	Treehugger	−0.615	0
		Nerd	−0.474	0
		Fatherlander	−0.582	0.0000001
Average emotionality	Spiritualism	Nerd	−0.010	0.0000259
		Treehugger	−0.0077	0.003112

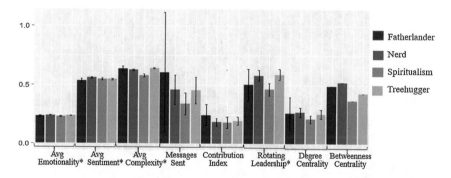

Fig. 2 Text and network metrics for the *Alternative Realities* macro-category (significance differences marked by asterisk)

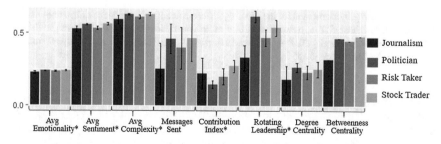

Fig. 3 Text and network metrics for the *Personality* macro-category (significance differences marked by asterisk)

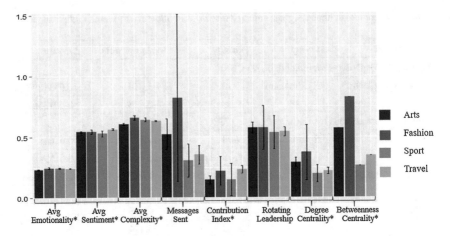

Fig. 4 Text and network metrics for the *Recreation* macro-category (significance differences marked by asterisk)

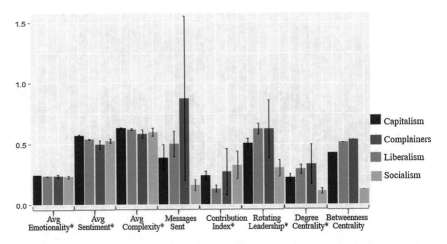

Fig. 5 Text and network metrics for the *Ideology* macro-category (significance differences marked by asterisk). *In these charts, Betweenness Centrality was divided by 500,000, Messages Sent by 200, Average Complexity by 10, Rotating Leadership by 100, and Degree Centrality by 100 in order to compare the data in one graph

datasets, suggesting that they bring less new ideas that Nerds, Fatherlanders, and Treehuggers. These differences are significant (p ranges from 0 to 0.033). There are no other significant differences in the personal email inbox, which may be due to the limited sample size. In the Enron dataset, the Spiritualists rotate their positions of leadership the least, which means that their betweenness centrality rarely oscillates. This suggests that in comparison to the Nerds and Treehuggers, Spiritualists rarely change their positions in a group and either stay as group leaders or followers.

Nerds speak with the highest positivity out of all the other groups as they have the highest average sentiment. Spirituals also have the least variation in the positivity of their messages, as demonstrated by their low Average Emotionality. All of these differences are significant as well (p ranges from 0 to 0.04).

In Table 2, for the *Recreation* macro-category, there is only one similarity between the Enron and email datasets. The Arts tribe has a lower average complexity than the Sports tribe, meaning they bring fewer ideas to the table ($p = 0.0002$, 0.008). There were many differences between the data as well. Primarily, in the average complexity metric, the Fashion tribe had the lowest complexity in the personal email inbox, and it had the highest average complexity in the Enron data. The same was true for the average emotionality metric. In the private emails, there was relatively low variation in the positivity of emails coming from the Fashion tribe, but there was a relatively high variation in the Enron data. There were only significant differences in average complexity and emotionality for the email data. In the Enron data, members of the Travel tribe seem to be the most central, as they have a higher degree and betweenness centrality than members of the Fashion and Arts tribes. Moreover, the Travel tribe contributes relatively more than the Arts tribe, with a higher Contribution

Table 2 Differences in honest signals for *Recreation* tribes in email inbox and Enron data

Honest signal	Group	Group	Mean difference	*P*-value
Significant comparisons for email inbox, $N = 20$				
Average complexity	Sport	Arts	1.49	0.0002092
	Fashion	Arts	−3.37	0.0000166
		Sport	−4.86	0.0000002
		Travel	−3.54	0.0000081
Average emotionality	Fashion	Arts	−0.154	0.0001033
		Sport	−0.142	0.0003459
		Travel	−0.146	0.0001779
Significant comparisons for Enron data, $N = 1738$				
Degree centrality	Travel	Fashion	−15.34	0.0497952
		Arts	−7.334	0.0160727
Betweenness centrality	Travel	Fashion	−239167	0.0572116
		Arts	−111761	0.023589
Contribution index	Travel	Arts	0.0842	0.0011708
Average sentiment	Travel	Arts	0.0215	0.000007
		Sports	0.0333	0.0180155
Average complexity	Arts	Fashion	−0.525	0.0000107
		Sport	−0.373	0.0080805
		Travel	−0.279	0
Average emotionality	Arts	Fashion	−0.0134	0.0040415
		Travel	−0.0086	0.0000008

Table 3 Differences in honest signals among *Ideology* tribes in emails and Enron data

Honest signal	Group	Group	Mean difference	P-value
Significant comparisons for email inbox, $N = 20$				
Average complexity	Liberalism	Capitalism	1.237	0.0302374
Significant comparisons for Enron data, $N = 1738$				
Honest signal	Group	Group	Mean difference	P value
Degree centrality	Liberalism	Capitalism	6.83	0.0272564
		Socialism	17.93	0.0106698
Contribution index	Liberalism	Capitalism	−0.112	0.0000035
		Socialism	−0.195	0.0012411
Rotating leadership	Liberalism	Capitalism	11.748	0.0003462
	Socialism	Capitalism	−20.573	0.0161735
		Complainers	−32.243	0.0300972
		Liberalism	−32.322	0.0000184
Average sentiment	Capitalism	Complainers	0.07544	0.0000014
		Liberalism	0.0327	0
		Socialism	0.0458	0.0000446
Average complexity	Capitalism	Complainers	0.4923	0.0070199
		Liberalism	0.1172	0.0463337
		Socialism	0.3295	0.0114416
Average emotionality	Capitalism	Liberalism	0.0074	0.0000306
		Socialism	0.0127	0.0053641

Index. The Travel tribe also speaks in the most positive manner with the highest average sentiment out of all the recreational tribes. These results are also statistically significant (p ranges from 0.000002 to 0.0498).

In Table 3, the only significant comparison from the email inbox was that between the complexity of Liberalism and Capitalism tribes, where the Liberalism tribe displayed a wider vocabulary ($p = 0.03$). Surprisingly, the Enron data displayed a different trend, as the Capitalism tribe had a higher average complexity than Liberalism did ($p = 0.046$). Liberals seem to have the highest connectivity of all the Ideology tribes, as they have the highest degree centrality. However, they seem to communicate less relative to the content they receive, with lower contribution indices than the rest of the tribes. Moreover, the Socialism tribe seems to have the most changes in leadership positions, followed by the Capitalism, Liberalism, and Complainers tribes. The Capitalism tribe speaks most positively in its messages, with an average sentiment higher than the rest of the tribes. Moreover, it has the most oscillations in its sentiment, with the highest average emotionality. These results are all significant, with a maximum p value of 0.046 overall.

In Table 4, there are no significant differences in the private email inbox. However, there are some differences in the Enron data. The Stock-Trader tribe seems to be more

Table 4 Differences in honest signals among *Personality* tribes in Enron Data

Honest signal	Group	Group	Mean difference	P value
No significant differences in email inbox, $N = 20$				
Significant differences in Enron data, $N = 1738$				
Contribution index	Stock-trader	Politician	0.127	0.000008
Rotating leadership	Politician	Risk-taker	14.36	0.0015132
		Journalist	28.36	0.0001742
Average sentiment	Journalist	Politician	−0.03	0.0050642
		Stock-trader	−0.036	0.0037871
	Risk-taker	Politician	−0.0248	0.0001364
		Stock-trader	−0.0272	0.000232
Average complexity	Journalist	Politician	−0.361	0.0030641
		Stock-trader	−0.392	0.0019163
	Risk-taker	Politician	0.172	0.0241113
Average emotionality	Journalist	Politician	−0.0116	0.0111244
		Stock-trader	−0.025	0.0008865

productive to conversations than the Politician tribe, as it has a larger Contribution Index. However, the Politician tribe changes its position more in discussions than the Risk-Taker and Journalist tribes, with a high rotating leadership. The Journalists and Risk-Takers speak most positively in discussions, as they have the highest average sentiments. Moreover, the Journalists have the largest deviations in the positivity of their messages in comparison to the Politicians and Stock-traders, given that they have a high average emotionality.

Discussion and Implications

This paper's findings add to the theoretical and practical study of tribes. We illustrate the applicability of this concept also for the analysis of organizations, extending it from its main use to marketing. From an academic standpoint, this paper expands the use of *Tribefinder* to the email setting. Earlier work has been mainly focused on social network sites like Twitter [7]. Since email databases also behave like a social network [26], the same methodology could be applied there. Moreover, this paper utilizes a new tool developed by Gloor et al. [7] in order to identify tribes. This allows us to circumvent the traditional methods of identifying tribes, like focus groups and interviews [18]. These groups have their traits analyzed through honest signals [10, 31], which demonstrates that there are differences among the tribes that have impacts on communication habits. Finally, this paper furthers work done in the field of data-driven human resources management and decision making, which has been taking recent strides [4], by dividing email users into groups based on their social network behavior and word content.

This paper is of importance in a managerial sense as well. For many companies, tribes have emerged as a critical factor of their success, especially in marketing [3]. In this paper we illustrate the usefulness of this concept in HR management as well. Many human resource managers have begun to analyze the traits of their employees through emails [5], but division of their users based on their traits has seen limited use. The use of digital social networks in HR is important due to the ease of access and spreading of information in the modern-day internet [6].

Limitations and Future Work

This work clearly has some limitations. Primarily, workers do not only communicate through email and use messaging services such as Slack and social networks such as Facebook. It could be beneficial to also analyze these sources to identify if these results are generalizable. There are also other models that could be used to identify the tribes of certain users, and they could yield different, and potentially more accurate results. Finally, other honest signals could be used besides those in this paper. Average response time and nudges (the amount of emails one sends in order to get a reply from another) could be used.

Conclusion

This paper illustrates the usefulness of the tribe concept for HR analysis. It shows the use of *Tribefinder* in a different medium and framework. It analyzes the communication habits of people in organizations through the lens of emails, utilizing LTSMs and word embeddings, and places them into tribes that the user can flexibly create depending on the focus of analysis. Four macro-categories of tribes are employed: Alternative Realities, Ideologies, Recreation, and Personality. However, this system could easily be extended for instance to measure moral values of employees, or their attitudes toward risk by creating the appropriate tribes. By comparing the tribal affiliations with the "honest signals of communication", we illustrate the underlying traits of different groups of employees, thus providing valuable cues to managers about the characteristics of their employees. This paper is an early research, but it clearly demonstrates the power of this approach to discover the underlying individual attributes and behavioral characteristics of members of an organization otherwise not accessible.

Acknowledgements All procedures performed in studies involving human participants were in accordance with the ethical standards of the institutional and/or national research committee with the Helsinki declaration and its later amendments or comparable with ethical standards.

References

1. B. Cova, V. Cova, Tribal marketing: the tribalisation of society and its impact on the conduct of marketing. Eur. J. Mark. **36**(5/6), 595–620 (2002)
2. B. Cova, The postmodern explained to managers: implications for marketing. Bus. Horiz. **39**(6), 15–24 (1996)
3. P.A. Gloor, A.F. Colladon, Heart beats brain-measuring moral beliefs through E-mail analysis, in *Proceedings of the 9th International Conference on Collaborative Innovation Networks (COINs)*, Warsaw, Poland (2019)
4. S. Strohmeier, F. Piazza, Domain driven data mining in human resource management: a review of current research. Expert Syst. Appl. **40**(7), 2410–2420 (2013)
5. B. Marr, *Data-driven HR: How to Use Analytics and Metrics to Drive Performance* (Kogan Page Publishers, 2018)
6. T.L. Adams, S.A. Smith (eds.), *Electronic Tribes: The Virtual Worlds of Geeks, Gamers, Shamans, and Scammers* (University of Texas Press, 2009)
7. P. Gloor, A.F. Colladon, J.M. de Oliveira, P. Rovelli, Put your money where your mouth is: Using deep learning to identify consumer tribes from word usage. Int. J. Inform. Manage. (2019)
8. S. Hochreiter, J. Schmidhuber, Long short-term memory. Neural Comput. **9**(8), 1735–1780 (1997)
9. B. Klimt, Y. Yang, The enron corpus: A new dataset for email classification research, In *European Conference on Machine Learning* (Springer, Berlin, Heidelberg, 2004), pp. 217–226
10. P.A. Gloor, *Sociometrics and Human Relationships: Analyzing Social Networks to Manage Brands, Predict Trends, and Improve Organizational Performance* (Emerald Publishing Limited, 2017)
11. B. Cova, V. Cova, Tribal aspects of postmodern consumption research: the case of French in-line roller skaters. J. Consum. Behav. Int. Res. Rev. **1**(1), 67–76 (2001)
12. M. Holzweber, J. Mattsson, C. Standing, Entrepreneurial business development through building tribes. J. Strategic Market. **23**(7), 563–578 (2015)
13. J. Turner, M.A. Hogg, P.J. Oakes, S.D. Reicher, M.S. Wetherell, *Rediscovering the social group: A social categorization theory* (Oxford, UK: B. Blackwell, 1987))
14. M.J. Hornsey, Social identity theory and self-categorization theory: a historical review. Soc. Pers. Psychol. Compass **2**(1), 204–222 (2008)
15. N. Ellemers, P. Kortekaas, J.W. Ouwerkerk, Self-categorisation, commitment to the group and group self-esteem as related but distinct aspects of social identity. Eur. J. Soc. Psychol. **29**(2–3), 371–389 (1999)
16. T. Garry, A.J. Broderick, K. Lahiffe, Tribal motivation in sponsorship and its influence on sponsor relationship development and corporate identity. J. Market. Manage. **24**(9–10), 959–977 (2008)
17. M. Maffesoli, *The Time of the Tribes: The Decline of Individualism in Mass Society*, vol. 41 (Sage, 1995)
18. C. Mitchell, B.C. Imrie, Consumer tribes: membership, consumption and building loyalty. Asia Pacific J. Market. Logistics **23**(1), 39–56 (2011)
19. B. Cova, From marketing to societing: When the link is more important than the thing. Rethink. Market. Towards Crit. Market. Account. 64–83 (1999)
20. Z. Bauman, *Thinking Sociologically* (B. Blackwell, Oxford; Cambridge, Mass, 1990)
21. K. Hamilton, P. Hewer, Tribal mattering spaces: social-networking sites, celebrity affiliations, and tribal innovations. J. Market. Manage. **26**(3–4), 271–289 (2010)
22. L.T. Wright, B. Cova, S. Pace, Brand community of convenience products: new forms of customer empowerment–the case "my Nutella The Community". Eur. J. Market. (2006)
23. T.E. Murphy, S. Zandvakili, Data-and metrics-driven approach to human resource practices: using customers, employees, and financial metrics. Hum. Resour. Manage. **39**(1), 93–105 (2000)

24. E. Brynjolfsson, L.M. Hitt, H.H. Kim, *Strength in numbers: How does data-driven decision-making affect firm performance?* Available at SSRN 1819486 (2011)
25. R. Dealtry, E.A. Smith, Communities of competence: new resources in the workplace. J. Workplace Learn. (2005)
26. C. Bird, A. Gourley, P. Devanbu, M. Gertz, A. Swaminathan, Mining email social networks, in *Proceedings of the 2006 International Workshop on Mining Software Repositories* (ACM, 2006), (pp. 137–143)
27. L. Moutinho, P. Dionísio, C. Leal, Surf tribal behaviour: a sports marketing application. Market. Intell. Plann. **25**(7), 668–690 (2007)
28. D. Oliveira, J. Marcos, P.A. Gloor, *GalaxyScope: finding the "Truth of Tribes" on social media*, Collaborative Innovation Networks (Cham, Springer, 2018), pp. 153–164
29. K. Greff, R.K. Srivastava, J. Koutník, B.R. Steunebrink, J. Schmidhuber, LSTM: A search space odyssey. IEEE Trans. Neural Netw Learn. Syst. **28**(10), 2222–2232 (2016)
30. D.T. Vo, C.Y. Ock, Learning to classify short text from scientific documents using topic models with various types of knowledge. Expert Syst. Appl. **42**(3), 1684–1698 (2015)
31. A. Pentland, *Honest Signals: How They Shape Our World* (MIT press, 2010)
32. L.C. Freeman, Centrality in social networks conceptual clarification. Soc. Net. **1**(3), 215–239 (1978)
33. S. Wasserman, K. Faust, *Wasserman, Stanley, and Katherine Faust, social network analysis: methods and applications* (Cambridge University Press, New York, 1994), p. 1994
34. Y.H. Kidane, P.A. Gloor, Correlating temporal communication patterns of the Eclipse open source community with performance and creativity. Comput. Math. Organ. Theory **13**(1), 17–27 (2007)

Chapter 9
The Political Debate on Immigration in the Election Campaigns in Europe

Francesca Greco and Alessandro Polli

Abstract Migration has become an increasingly pressing topic on the national and European political agendas and in general public debate. The migratory phenomenon, as well as its humanitarian and health implications, are presented nowadays as a challenge for national and supranational governments which requires coordinated responses to ensure citizen security. During the election campaigns in the last three years, right-wing parties have largely depicted the right of freedom of movement as a risk factor, taking advantage of this issue for political propaganda. A significant part of the political debate takes place on social media, which has become the preferred platform for openly expressing political sentiment, including that considered politically incorrect. This study explores the political debate on immigration during the election campaigns of France and Italy over the last three years. More specifically, we perform Emotional Text Mining with the aim of identifying the sentiment surrounding immigration, and how immigrants are portrayed, in the online Twitter debate during the French presidential election (2017), the Italian general election (2018), and the Italian European elections (2019). Results were compared to identify the similarities and differences, and the effect on the election results that characterized two of the European Union's founding countries.

Introduction

Migration has become an increasingly pressing topic on the national and European political agendas and in general public debate. This topic is politically sensitive and challenging for national governments and the EU parliament, alike. It touches on both the question of residence and the right of freedom of movement within the EU, and

F. Greco (✉) · A. Polli
Sapienza University of Rome, 5 Piazzale Aldo Moro, Rome, Italy
e-mail: francesca.greco@uniroma1.it

A. Polli
e-mail: alessandro.polli@uniroma1.it

© Springer Nature Switzerland AG 2020
A. Przegalinska et al. (eds.), *Digital Transformation of Collaboration*,
Springer Proceedings in Complexity,
https://doi.org/10.1007/978-3-030-48993-9_9

humanitarian and health implications. Public debate on immigration took on particular emphasis during the general and European elections in Italy, and the presidential elections in France. In all three occasions, the focus of the electoral consensus was directed onto specific political proposals which represented immigration as a potential risk factor for EU citizens [2, 4]. These proposals have been associated with a strengthening of nationalism and has put into question EU membership. Currently, the issue of immigration reveals the difficulty the member states have in finding a common strategy on the management of new entrants. This, in turn, has heightened the debate on membership of the European Union itself, even among its founding countries.

In recent years, in Italy and France, right-wing political parties have largely focused their propaganda on emotive, national identity issues, including leaving the European Union and fighting immigration. In contrast to past modes of communication, the latest form, social media, has become massively popular and now represents the preferred platform for expressing politically incorrect sentiments.

The reason why political propaganda is flooding social media platforms is that they increasingly fulfil the function of communication, not only enabling millions of users to share information daily but also funnelling citizens' comments, opinions, and feelings on a wide range of topics. Accordingly, social media and social networking sites, such as Facebook and Twitter, have begun to take on a growing role in real-world politics [5–7], shifting political communication towards the new digital media. Not surprisingly, one of the most important features of the recent public debate on immigration is that it mainly took place on social media platforms, as highlighted by a study carried out during the latest French presidential campaign [21].

As a result of this shift, a growing number of social media analysis techniques have been developed to explore a range of politically orientated topics, such as the organization of demonstrations and actions of revolt, the establishment of and participation in social movements and political parties and electoral campaigns (e.g. [8, 9, 18, 21, 23, 29]).

This study explores the political debate on immigration during the election campaigns in France and Italy over the last three years. More specifically, we perform a quantitative study aiming to identify the sentiment surrounding immigration, and how immigrants are portrayed, in the online Twitter debate during the France presidential election (2017), the Italian general election (2018), and the Italian European elections (2019). We compare the results to identify the similarities and differences that characterized two of the founding member states of the European Union.

Emotional Text Mining

Sentiment analysis is a field of study that analyses people's opinions, sentiments, evaluations, appraisals, attitudes, and emotions towards entities. It is also called opinion mining, since, frequently, the sentiment is considered a personal belief, or judgment, which is not founded on rational reasoning, but on a subjective emotion. The use of a text mining approach to classify the sentiment of a text has been largely discussed in the literature (e.g. [1, 3, 12, 14, 27]). Most methods are based on a top-down approach where an a priori coding procedure of terms, or text, is performed focusing on the manifest content of the word. Emotional Text Mining (ETM) is a particular kind of sentiment analysis based on a socio-constructivist approach and a psychodynamic model, which allows for the identification of the elements dictating people's interactions, behaviour, attitudes, expectations, and communication [15, 19].

According to the semiotic approach of the analysis of textual data, ETM allows for social profiling to be carried out. This has already been applied in different fields ranging from political debate [16–18, 20, 21], to professional training [11], brain structure [25], brand management [19], and to the impact of the law on society (e.g. [10, 15, 22]).

While mental functioning proceeds from the semiotic level to the semantic one in generating communication, the statistical procedure simulates the inverse process of mental functioning, from the semantic level to the semiotic one. For this reason, ETM performs a sequence of synthesis procedures, from the reduction of the type to lemma and the selection of the keywords, to the clustering and factorial analysis. This allows for the identification the semiotic level (the symbolic matrix) starting from the semantic one (the word co-occurrence).

In order to perform ETM, we collect all the messages into a corpus and calculate the lexical indices (token, TTR, hapax percentage) in order to check whether it is possible to statistically process the data. Data are cleaned and pre-processed [24] and keywords selected, filtering out the terms used to select the messages and those belonging to the low rank of frequency [19]. On the tweets per keyword matrix, we perform a cluster analysis with a bisecting k-means algorithm based on cosine similarity [28, 30] limited to 20 partitions, excluding all the tweets that did not have at least two keywords co-occurrence. To choose the optimal solution, we calculate the Calinski-Harabasz, the Davies-Bouldin, and the intraclass correlation coefficient (ρ) indices. Then, we perform a correspondence analysis [26] on the cluster per keywords matrix, and we calculate the sentiment according to the number of messages classified in the cluster and its interpretation.

The Immigration Debate During the Election Campaign

In order to explore the public perception of immigration in Twitter communications during the election campaigns, we scraped all the messages from the Twitter repository, containing the words "immigrant/s" and "immigrations" during the election campaign. The data extraction was carried out with the twitterR package of R Statistics [13].

French Presidential Election

Data collection was conducted during the period immediately preceding the first round of the French presidential elections (April 23, 2017). We scraped all the messages written in French produced by the Twitter repository from April 10 to April 22, 2017. The sample of 111,767 tweets was made up of 77.7% retweets and resulted in a large corpus (token = 2,154,194; TTR = 0.01; Hapax percentage = 40.4). Results of the cluster analysis show that the 625 keywords selected allow for the classification of 90% of the tweets. The clustering validation measures show that the optimal solution is seven clusters. In Fig. 9.1, we can discern the emotional map of immigration emerging from the French messages. It shows how the clusters are placed in the symbolic space produced by the first three factors, explaining 61% of the inertia.

The seven clusters are of different sizes and reflect different representations of immigration that correspond to three different sentiments: positive, negative for the community, and negative for immigrants. The first cluster (9.9%) reflects the reaction of public opinion to the proposal of hosting immigrants in historical buildings, suggested by Mélenchon. Immigrants seem to be perceived as undesirable guests or

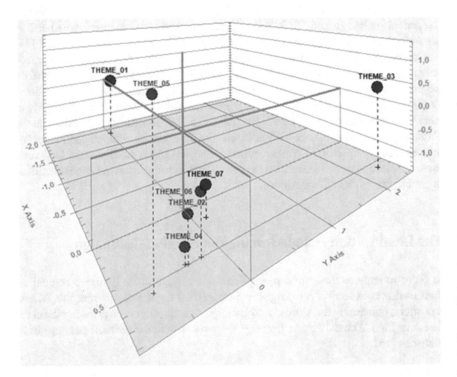

Fig. 9.1 Emotional map of immigration emerging from the French messages (the factorial space is set by the first three factors) [21]

squatters. The second cluster (17.7%) highlights the association between immigration and security. Immigrants are represented as terrorists and aggressors, and a multitude of religious beliefs within a population are perceived as an obstruction to cohabitation. Immigration is represented as a risk factor for security because it favours terrorism, and this association seems to be driven by the speeches of the *National Front* leader. The third cluster (3.6%) refers to human trafficking that changes the sense of mobility from a voluntary choice to an involuntary fate, transforming a journey of hope into a nightmare and the immigrants into slaves. The fourth cluster (21.2%) reflects the EU reception policy, in which receiving, hosting, and integrating immigrants are some of the activities in need of an appeal for solidarity from French citizens. In the fifth cluster (17.7%), immigrants are perceived as invaders, which is a theme that often runs through the *National Front* leader's speeches. Among the words of this cluster, there are words such as border, stop, wave, control (frontière, arrêter, vague, maîtriser) which are combined with insults and swearing, highlighting a significant level of anger. The sixth cluster (21.1%) represents immigrants as bringing prestige to the country, for example, celebrity sportsmen and women. There are the first names, or surnames, of famous football players and gymnastic champions among the words of this cluster that are associated with goal and cup, and also diversity. Here, diversity seems to be a positive value that distinguishes sports players. Lastly, the seventh cluster (8.8%) reflects the dangers entailed in the immigration journey and how people's lives are put in jeopardy. Only a proportion of the immigrants who sail across the Mediterranean will land. The others will never arrive nor return.

By the cluster's interpretation, we detected seven different representations of immigrants that correspond to three different sentiments: positive (42%), negative for the community (45%), and negative for immigrants (12%). We have considered as negative the representation of immigrants as squatters, invaders, terrorists, trafficked slaves, and immigration victims, and as positive, the sporting celebrities and the EU solidarity target. Among the negative clusters, we distinguished negativity according to the direction of the activity: squatters, terrorists, and invaders are negative for the community and trafficked slaves and fatalities linked to immigration are negatives for the immigrants themselves. In Fig. 9.2, the dimensions of the three sentiments are displayed.

Fig. 9.2 French sentiment on immigration during the presidential election campaign [21]

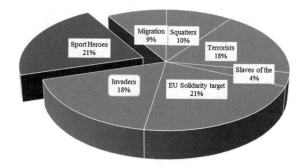

It is interesting to note that during the presidential elections in France, the *National Front* leader's political propaganda largely depicted the theme of freedom of movement as a risk factor, focusing the political campaign on topics such as border closure and exiting from the EU. In the second round, she lost the election, taking 34% of votes. This percentage almost corresponds to the sum of the clusters in which immigrants are represented as terrorists and invaders (35.4%), the main topics of the leader's political discourse.

Italian General Election

Data collection was performed almost two months before the Italian general election (March 4, 2018). We scraped all the messages written in Italian produced from January 16 to January 25, 2018, from the Twitter repository. The sample of 41,157 tweets was made up of 84% retweets and resulted in a large corpus (token = 738,897; TTR = 0.03; Hapax percentage = 40.3). Results of the cluster analysis show that the 523 keywords selected allow for the classification of 91% of the messages. The clustering validation measures show that the optimal solution is six clusters. In Fig. 9.3, we can discern the emotional map of immigration emerging from the

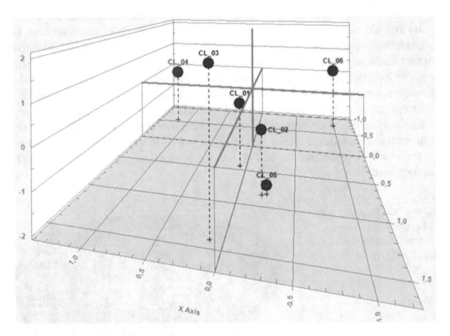

Fig. 9.3 Emotional map of immigration emerging from the Italian messages (the factorial space is set by the first three factors) [16]

Italian messages. It shows how the clusters are placed in the symbolic space produced by the first three factors, explaining 70% of the inertia.

Despite the different tones from those prevailing in the French debate, due to the absence of swear words, the Italian debate on Twitter mainly presents negative aspects. The first cluster represents immigration as a dangerous invasion of the country, which the authorities endeavour to stem through a political programme designed to regulate the flow of immigration. This theme shows great similarities to the cluster of *invaders* identified in the French messages. In both cases, the two clusters classify a very significant percentage of texts. The second cluster refers to the theme of *new slaves*, meaning that in the absence of protection against exploitation of immigrants, the immigration process is characterized by an aspect of illegality. Unlike the cluster on *Illegal Workers* (cluster 6), this theme seems to refer to an aspect more directly connected to the political debate that characterizes the electoral campaign. The third cluster represents immigrants as violent towards women, categorizing specific cultures, such as the Islamic one, as uncontrollable and conveying values contrary to those of the host country. The fourth cluster represents immigrants as a cost to the community that requires government decision making and action to take into account the economic repercussions to the country. While in the French campaign the theme of solidarity affirms a positive sentiment, in the Italian one immigrants are represented more as a social burden that has to be managed, which evokes a neutral sentiment. The fifth group refers to a specific event in which young legal immigrants attacked a police station in Sweden. Like the reactions to the statements of the extreme-left candidate in France, the texts focus on the news. However, unlike the French messages, the issue of security risks related to legal immigration emerges from Italian texts with feelings of anger. However, the Italian texts seem to be more politically correct and appear to be less aggressive than the French ones. Finally, the sixth cluster recalls the theme of the exploitation of illegal immigrants, who are employed to carry out the humblest jobs and who are forced to live in hiding.

The sentiment in the Italian corpus lacks positivity and only 12% of the classified messages are neutral. Tweets are mostly negative (88%) and negativity can be distinguished as negative for the community (33%), negative for immigrants (39%), and gender negativity (16%). The type of negativity is almost the same in the French communications, but it lacks the gender characterization.

Although globally there is a more negative sentiment in Italy than in France, the negative sentiment seems to focus more on personal aspects (negative for immigrant and gender negativity = 55%) rather than community ones. This seems to suggest that Italian culture is more sensitive to individual elements. Moreover, the different geopolitical conditions that characterize the two countries probably involve less positive sentiment regarding the need to manage the problem of immigration costs at the European level (Fig. 9.4).

Fig. 9.4 Italian sentiment on immigration during the general election campaign

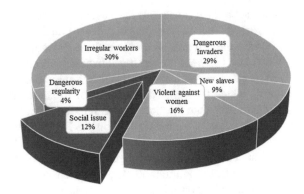

Italian European Election

The European elections were a crucial test of the general election results of 2018, particularly for the ruling parties: *Lega* and *Movimento Cinque Stelle* (M5S). Although the campaigns of the Italian political parties emphasized domestic issues, a main topic of the political debate called into question the immigration issue, criticizing the EU's governance. The general dissatisfaction of Eurozone rules and the need for greater solidarity divided the political parties into two camps: those in favour of a revision of the Dublin regulation for a common immigration policy based on solidarity with fair redistribution and division of responsibilities among the EU countries, and those emphasizing the need to hinder immigration by protecting external borders, increasing effective repatriations and opposing the redistribution of immigrants.

As in the Italian political campaign of 2018, data collection was performed more than two months before the Italian vote (May 26, 2019). We scraped all the messages written in Italian produced from March 19 to April 14, 2019, from the Twitter repository. The sample of 96,681 tweets was made up of 75% retweets and resulted in a large corpus (token = 1,851,083; TTR = 0.02; Hapax percentage = 61.7). Results of the cluster analysis showed that the 870 keywords selected allowed for the classification of 93% of the messages. The clustering validation measures showed that the optimal solution was five clusters. In Fig. 9.5, we can discern the emotional map of immigration emerging from the Italian messages. It shows how the clusters are placed in the symbolic space produced by the first three factors, explaining 83% of the inertia.

The five representations of immigrants are of different proportions and correspond to two sentiments: positive (16%) and negative (84%) (Fig. 9.6). The first cluster (21.9%) reflects the reaction of public opinion to the right-wing parties' political manipulation of the immigration issue. Italian people seem to perceive the left-wing parties' proposal of a common immigration policy based on solidarity as a betrayal, while they consider favourably the *zero-tolerance* attitude of the *Lega* leader. In this cluster, illegal immigrants are considered as dangerous criminals, and the comments

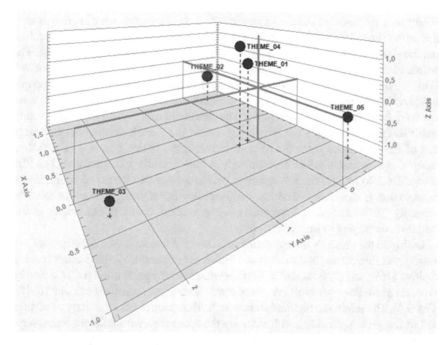

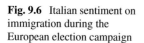

Fig. 9.5 Emotional map of immigration emerging from the Italian messages (the factorial space is set by the first three factors)

Fig. 9.6 Italian sentiment on immigration during the European election campaign

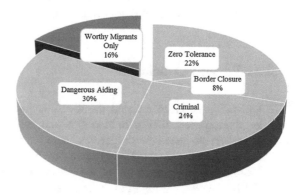

are particularly racist, while the Italian population are represented as victims of the left-wing parties' policy of aiding and abetting illegal immigration.

The second cluster (16.4%) identifies the worthy immigrants that can be welcomed into Italy. Hence, while generalization seems to induce people to represent immigrants negatively, the possibility of distinguishing one person from another highlights the positivity relating to specific people. In fact, immigrants who work and cooperate honestly with Italians are considered worthy of inclusion. In the third cluster (8.2%), immigrants are perceived as dangerous invaders. This representation is often associated with the need for border closure, in line with the *Lega* leader's political speeches, and those of the *National Front leader* during the French presidential campaign. The instrumental use of specific crime news increased the perception of insecurity about living among immigrants. In fact, this cluster contains names referring to people and places reported in the news which concern crimes in which immigrants were involved, both as victims or offenders. The fourth cluster (23.5%) represents immigrants as criminals. While the stereotype during the French presidential campaign was of the immigrant offender being a terrorist, the Italian one was of a violent criminal attempting to attack personal safety, both sexually and otherwise. The fifth cluster (30.1%) reflects the perception in public opinion that the act of aiding illegal immigration is dangerous. The hospitality policy of left-wing parties towards the allegedly *"spoiled"* illegal immigrants is represented as aiding and abetting, and therefore, damaging to Italian citizens.

Owing to the cluster's interpretation, we detected five separate representations of immigrants that correspond to only two different sentiments. We have considered as positive (16%) only the second cluster, in which worthy legal immigrants would be welcomed into Italy, while all the others were classified as negative sentiments (84%) (Fig. 9.6). The sentiment on immigration in the European elections in Italy confirmed that of the general election, with virtually the same negative sentiment percentage. Nonetheless, the overall winner of the elections was the *Lega*, the right-wing ruling party, who focused its propaganda on the need to close the border, reinforce control, and improve repatriations. The *Lega*, who secured only 6% of votes in the 2014 European elections, tripled its consensus in the 2018 general election and obtained 34% of the votes in the 2019 European election.

Discussion and Conclusion

The application of ETM enabled the identification of several points of contact between the representation of immigration and the related debates in France and Italy. However, alongside the similarities, there are some significant differences.

The first general consideration refers to the time window for collecting textual data. Not surprisingly, the width of the chosen window affects the results. More specifically, the closer we get to the electoral deadline, the more the texts seem to take a specific structure and tone. The tone of the tweets shared in France appear more visceral and direct, while the texts collected in Italy seem to be characterized by a more reflective and politically correct tone. This discrepancy likely reflects the window of time chosen for the collection of textual data, rather than other factors, such as a possible difference in cultural traits that could affect the authors' sentiment towards the issue of immigration.

The way the issue of immigration was manipulated by the French and Italian right-wing parties appears to have been effective in terms of political communication. If we compare the voting percentages of the right-wing parties during the elections with the sentiment of the tweets, we note a substantial correlation between the positions expressed in the debate on social media and the election results. Nevertheless, it is still not clear how to implement sentiment analysis to forecast election results in a robust way.

An interesting result obtained by applying ETM is the similarity observed between French and Italian sentiment, which were both essentially negative towards immigration, albeit with some significant differences. Another common feature is the attitude of perceiving immigrants as at-risk, weak, and in need of protection. This supposed condition of vulnerability is due to two factors: a) how the immigrants arrive in Europe, and b) their future position in the labour market of the host country, where they risk being exploited as low-cost labour, in conditions verging on slavery.

A common theme in the debate on social media concerns solidarity having a dual interpretation. On the one hand, the solidarity debate at the level of European policies requires Europe to deal with the issue by adopting more supportive policies for first-entrant countries. On the other hand, solidarity refers to social solidarity towards immigrants. In the French debate on social media, the sentiment of social solidarity is essentially positive, while it is neutral in the Italian tweets, which is more focused on the costs of welcoming immigrants.

Finally, typically national prejudices and stereotypes emerge in this discussion. In France, the immigration debate often equates the immigrant to terrorists. However, in Italy, it expresses concern about both the alleged violent attitude of immigrants towards women, and the consequences for children (immigrant and resident alike) due to the possible onset of integration problems.

In conclusion, the research allowed for the detection of a substantial correlation between the sentiment expressed on a theme used by a political party during the election campaign and the voting percentages expressed for the same party in the election. This result is of significant interest because we have reduced the cost of the research using social media data, thanks to the possibility of analysing the political debates using multivariate statistical methodologies, in comparison to undertaking three expensive surveys.

Although the last three electoral rounds held in Italy and France have produced relatively equivalent results, we are unable to certify the robustness of a forecasting method centred on sentiment as a leading indicator. Therefore, future research will concern the application of text mining techniques and the elaboration of composite indicators aimed at this purpose.

References

1. S. Balbi, M. Misuraca, G. Scepi, Combining different evaluation systems on social media for measuring user satisfaction. Inf. Process. Manag. **54**(4), 674–685 (2018)
2. M. Binotto, M. Bruno, V. Lai, *Tracciare confini L'immigrazione nei media italiani* (Franco Angeli, Milano, 2016)
3. J. Bollen, H. Mao, X. Zeng, Twitter mood predicts the stock market. J. Comput. Sci. **2**(1), 1–8 (2011)

4. D. Carzo (ed.), *Narrare l'Altro Pratiche discorsive sull'immigrazione* (Aracne, Roma, 2011)
5. M. Castells, Communication power and counter-power in the network society. Int. J. Commun. **1**, 238–266 (2007)
6. C. Cepernich, E. Novelli, Sfumature del razionale La comunicazione politica emozionale nell'ecosistema ibrido dei media. Comunicazione politica **1**, 13–30 (2018)
7. A. Ceron, L. Curini, S.M. Iacus, *Social Media e Sentiment Analysis L'evoluzione dei fenomeni sociali attraverso la Rete* (Springer, Milano, 2013)
8. A. Ceron, L. Curini, S.M. Iacus, iSA a fast scalable and accurate algorithm for sentiment analysis of social media content. Inf. Sci. **367**, 105–124 (2016)
9. A. Ceron, L. Curini, S.M. Iacus et al., Every tweet counts? How sentiment analysis of social media can improve our knowledge of citizens' political preferences with an application to Italy and France. New Media Soc. **16**(2), 340–358 (2014)
10. B. Cordella, F. Greco, K. Carlini et al., Infertilita e procreazione assistita: evoluzione legislativa e culturale in Italia. Rassegna di Psicologia **35**(3), 45–56 (2018). https://doi.org/10.4458/141 5-04
11. B. Cordella, F. Greco, P. Meoli et al., Is the educational culture in Italian Universities effective? A case study, in *JADT' 18: Proceedings of the 14th International Conference on Statistical Analysis of Textual Data*, ed. by D.F. Iezzi, L. Celardo, M. Misuraca (Universitalia, Rome, 2018), pp. 157–164
12. A. Fronzetti Colladon, The semantic brand score. J. Bus. Res. **88**, 150–160 (2018)
13. J. Gentry, *R Based Twitter Client*. R package version 1.1.9 (2016)
14. P.A. Gloor, *Sociometrics and Human Relationships: Analyzing Social Networks to Manage Brands Predict Trends and Improve Organizational Performance* (Emerald Publishing Limited, London, 2017)
15. F. Greco, *Integrare la disabilità Una metodologia interdisciplinare per leggere il cambiamento culturale* (Franco Angeli, Milano, 2016)
16. F. Greco, Il dibattito sulla migrazione in campagna elettorale: Confronto tra il caso francese e italiano. Culture e Studi nel Sociale **4**(2), 205–213 (2019)
17. F. Greco, L.S. Alaimo, L. Celardo, Brexit and Twitter: the voice of people, in *JADT' 18: Proceedings of the 14th International Conference on Statistical Analysis of Textual Data*, ed. by D.F. Iezzi, L. Celardo, M. Misuraca (Universitalia, Rome, 2018), pp. 327–334
18. F. Greco, L. Celardo, L.S. Alaimo, Brexit in Italy: text mining of social media, in *Book of Short Papers SIS 2018*, ed. by A. Abbruzzom, D. Piacentino, M. Chiodi, E. Brentari (Pearson, Milano, 2018), pp. 767–772
19. F. Greco, A. Polli, Emotional text mining: customer profiling in brand management. Int. J. Inf. Manag. (2019). https://doi.org/10.1016/j.ijinfomgt.2019.04.007
20. F. Greco, A. Polli, Vaccines in Italy: the emotional text mining of social media. RIEDS **73**(1), 89–98 (2019)
21. F. Greco, D. Maschietti, A. Polli, Emotional text mining of social networks: the French pre-electoral sentiment on migration. RIEDS **71**(2), 125–136 (2017)
22. F. Greco, S. Monaco, M. Di Tran et al., Emotional text mining and health psychology: the culture of organ donation in Spain, in *ASA Conference 2019—Book od Short Papers Statistics for Health and Well-being University of Brescia September 25–27 2019*, ed. by M. Carpita, L. Fabbris (CLEUP, Padova, 2019), pp. 125–129
23. D. Hopkins, G. King, A method of automated nonparametric content analysis for social science. Am. J. Pol. Sci. **54**(1), 229–247 (2010)
24. F. Lancia, *User's Manual: Tools for Text Analysis*. T-Lab version Plus 2018 (2018)
25. D. Laricchiuta, F. Greco, F. Piras et al., "The grief that doesn't speak" Text mining and brain structure, in *JADT' 18: Proceedings of the 14th International Conference on Statistical Analysis of Textual Data*, ed. by D.F. Iezzi, L. Celardo, M. Misuraca (Universitalia, Rome, 2018), pp. 419–427
26. L. Lebart, A. Salem, *Statistique Textuelle* (Dunod, Paris, 1994)
27. B. Liu, *Sentiment Analysis Mining Opinions Sentiments and Emotions* (Morgan & Claypool, 2012)

28. S.M. Savaresi, D.L. Boley, A comparative analysis on the bisecting k-means and the PDDP clustering algorithms. Intell. Data Anal. **8**(4), 345–362 (2004)
29. H. Schoen, D. Gayo-Avello, P. Metaxas et al., The power of prediction with social media. Internet Res **23**(5), 528–543 (2013)
30. M. Steinbach, G. Karypis, V. Kumar, A comparison of document clustering techniques, in *KDD Workshop on Text Mining*, vol. 400 (2000), pp. 525–526

Chapter 10
Brand Intelligence Analytics

Andrea Fronzetti Colladon and Francesca Grippa

> *"In God we trust. All others must bring data".*
> W. Edwards Deming.

Abstract Leveraging the power of big data represents an opportunity for brand managers to reveal patterns and trends in consumer perceptions, while monitoring positive or negative associations of the brand with desired topics. This chapter describes the functionalities of the SBS Brand Intelligence (SBS BI) app, which has been designed to assess brand importance and provide brand analytics through the analysis of (big) textual data. To better describe the SBS BI's functionalities, we present a case study focused on the 2020 US Democratic Presidential Primaries. We downloaded 50,000 online articles from the Event Registry database, which contains both mainstream and blog news collected from around the world. These online news articles were transformed into networks of co-occurring words and analyzed by combining methods and tools from social network analysis and text mining.

Introduction: A Brand Intelligence Framework

In this paper we describe a new dashboard and web app to assess brand image and importance through the analysis of textual data and using the composite indicator known as Semantic Brand Score (SBS) [1, 2]. The predictive power of the SBS and its three dimensions, that is, prevalence, diversity and connectivity, has been demonstrated in various settings, including tourism management and political forecasting [1, 3].

A. Fronzetti Colladon (✉)
Department of Engineering, University of Perugia, Perugia, Italy
e-mail: andrea.fronzetticolladon@unipg.it

F. Grippa
Northeastern University, Boston, MA, USA
e-mail: f.grippa@northeastern.edu

© Springer Nature Switzerland AG 2020
A. Przegalinska et al. (eds.), *Digital Transformation of Collaboration*,
Springer Proceedings in Complexity,
https://doi.org/10.1007/978-3-030-48993-9_10

Different from traditional measures, the SBS has the benefit of not relying on surveys—which are usually subject to different biases (e.g. [4–6]). The analysis is not constrained by small samples or by the fact that interviewees know that they are being observed. A set of texts could represent the expressions of an entire population, as for example all the news articles about Greta Thunberg. The SBS can be calculated on any source of text, including emails, tweets and posts on social media. The goal is to take the expressions of people (e.g. journalists, consumers, CEOs, politicians, citizens) from the places where they normally appear. This is aligned with previous research which proposed software and algorithms to mine the web, identify trends and measure the popularity of people and brands [7, 8].

In this chapter, we describe how the SBS components can be fully translated into reports available to brand managers and digital marketing professionals. In order to demonstrate the benefits of the SBS BI app and describe some of the reports it generates, we apply the framework to the case of the 2020 US Democratic Presidential Primaries, by mining 50,000 online news articles and combining methods and tools of social network analysis and text mining.

In addition to the calculation of the SBS, the analysis we conduct is based on topic modeling, sentiment analysis and the study of word co-occurrences—which help reveal patterns and trends in consumer perceptions, identifying positive, neutral or negative associations of the brand with other topics [9]. The association between different concepts used in an online discourse to describe a brand can help marketing managers discern the perceived relationships among brands, as well as their positioning in the customers' mind.

The Semantic Brand Score

The Semantic Brand Score (SBS) [2] is a novel measure of brand importance, which is at the core of the analytics we describe in this chapter. It was designed to assess the importance of one or more brands, considering dynamic longitudinal trends using data from multiple online sources and different contexts. It is a measure suitable for the analysis of (big) textual data across cultural systems and languages. The SBS conceptualization was partially inspired by well-known brand equity models and by the constructs of brand image and brand awareness [10].

The concept of "brand" is very flexible and the SBS can be calculated for any word, or set of words, in a corpus. By "brand" one could intend the name of a politician, or multiple keywords representing a concept (e.g. the concept of "innovation" or a corporate core value). The measure was used to evaluate the transition dynamics that occur when a new brand replaces an old one [2], to evaluate the positioning of competitors, to forecast elections from the analysis of online news [1] and to predict trends of museums visitors based on tourists' discourse on social media [3].

The SBS has three dimensions: prevalence, diversity and connectivity. Prevalence measures the frequency of use of a brand name, that is, the number of times a brand is directly mentioned—which can be considered a proxy of brand awareness and recall.

Diversity measures the heterogeneity of the words associated with a brand, that is, the richness of its lexical embedding. Connectivity represents a brand's ability to bridge connections between other words, which can represent concepts or discourse topics. The sum of these three indicators measures brand importance. The metric is fully described by the work of Fronzetti Colladon [2].

Textual Brand Image

When designing and evaluating brand-building programs, marketing managers should assess and measure brand image and brand equity by using other brands as benchmark, based on similarity of their positioning strategies. At the same time, many brand managers face the challenge to identify and visualize the correct measures of brand strength to complement financial measures with brand asset measures [11]. Building a strong brand image requires the adoption of a comprehensive measurement system able to validate brand-building initiatives and continuously monitor the impact on customer perception. An emerging method to assess and build individual brand image is the study of the words used to describe a brand.

Some scholars are adopting the theory of memetics to develop prediction tools to assess the spread of innovations [12] or to understand how concepts and brands are positioned in the minds of consumers. As noted by Marsden [13], how a brand is positioned in the associative networks of memory can be used to describe the meaning of that idea for customers. Techniques such as memetic analysis and use of brand mapping [14] allow marketing managers to assess how brands are positioned in the minds of consumers and whether these associations are positive, negative or neutral. These insights will support a better positioning of brands to fit with the consumers' mindset.

Measuring brand similarity is useful when selecting the most appropriate brand name, or to understand how a brand resemblance with another could impact brand loyalty and price sensitivity. Measuring brand similarity is also key when assessing how complete and comprehensive is the information provided to customers [15].

Traditional methods are usually based on surveys and use aggregated judgments made by potential customers [16]. Looking at the association between concepts describing a brand can help identify the perceived and psychological relationships among brands, their relative positioning and strategic differentiation.

Content analysis and topic modeling methods offer insights in the main topics discussed online, providing a set of keywords, and their connections [16, 17]. In this context, sentiment analysis of online data (e.g. news, reviews, blog entries) also comes to help and measures users' emotions and users' polarity toward a specific event, public figure or brand.

Recent studies have mined large-scale, consumer-generated online data to understand consumers' top-of-mind associative network of products [18] by converting them into quantifiable perceptual associations and similarities between brands. Others have gone beyond the mere occurrence of terms in online data and assessed

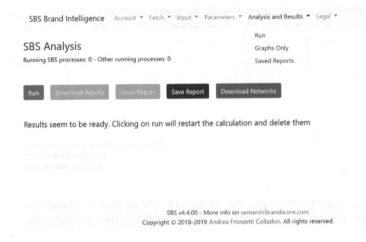

Fig. 10.1 SBS brand intelligence app

the proximity or similarity between terms using the frequency of their co-occurrence within a text [19]. Gloor and colleagues visualized social networks as Cybermaps and used metrics such as betweenness centrality and sentiment to evaluate the popularity of brands and famous people [7, 8].

The SBS BI Web App

The SBS Brand Intelligence (SBS BI) app has been developed to support the assessment of brand importance and the study of brand image and characteristics, through the analysis of (big) textual data.[1] This section describes the app's main components (version 4.5.10) and the analytical reports it generates.

As shown in Fig. 10.1, the app has several menus, starting with the option to upload and analyze any text file that is available to the user. The app has also modules that allow the fetching of online news and tweets. The fetching modules use the Twitter API[2] and the Event Registry API [20] in order to collect data. A dedicated option gives users the opportunity to connect to the Telpress[3] platform, for the collection of news and the download of data which perfectly integrates with the SBS BI app. After uploading a csv file, the user is expected to set a number of parameters—such as the language and time intervals of the analysis, the word co-occurrence range and the minimum co-occurrence threshold for network filtering (see [2] for more details).

---

[1] The SBS BI web app is distributed as Software-as-a-Service, and access can be requested for research purposes. Web address: https://bi.semanticbrandscore.com. Conceptualized and developed by Andrea Fronzetti Colladon (Copyright © 2018–2020).

[2] https://developer.twitter.com/en/docs/api-reference-index.

[3] http://www.telpress.com/.

The last step consists of running the core module, which will calculate the SBS and the other measures described in Sect. 3.2.

Text Preprocessing

A preliminary step before the calculation of all metrics is the preprocessing of uploaded texts, starting with removing web addresses, punctuation, stop-words and special characters. Documents are subsequently tokenized and words are converted to lowercase. Word affixes are removed through the snowball stemmer included in the NLTK Python package [21].

After the preprocessing phase, documents are transformed into undirected networks, based on word co-occurrences. In these networks, nodes represent words, and links among them are weighted based on co-occurrence frequencies. This serves to the calculation of the SBS. SBS BI gives users the option to download networks in the Pajek file format [22].

Calculation of the Semantic Brand Score

The SBS is the metric at the core of our analytics. Its dimensions of diversity and connectivity are calculated through the metrics of degree and weighted betweenness centrality [1, 2]. The traditional degree centrality metric can be adjusted [23] to value more the connections to low-degree nodes:

$$Diversity_i = \sum_{\substack{j=1 \\ j \neq i}}^{N} \log_{10} \frac{N-1}{g_j} I_{(w_{ij}>0)}$$

In the formula, $Diversity_i$ is the diversity of node i, N is the total number of nodes in the network, g_j is the degree of node j, and $I_{(w_{ij}>0)}$ is the indicator function which equals 1 if the edge connecting node i to node j exists, and 0 otherwise. We assume $w_{ij} = 0$ for unconnected nodes. The idea behind this adjustment is that associations of a brand are more distinctive if they occur with words having fewer connections. Several other variations of this metric are possible [23].

Prevalence is the count of word frequencies. Each measure is subsequently standardized, considering all the words in the network, by subtracting the mean to individual scores and dividing by the standard deviation. Standardized scores are added up to calculate the SBS. Other standardization techniques are also implemented by the app–such as min–max normalization or standardization obtained by subtracting the median and dividing by the interquartile range. Raw and standardized scores are provided as output.

Lastly, the SBS can also be calculated attributing different weights to different text documents. For example, the analyst might want to consider as more important an article published by The New York Times than one published by The Onion newspaper. Weights can be determined by the user and uploaded into the system. One possible approach for the determination of weights of online news is to refer to the Alexa[4] ranking of their sources. Other factors could also impact brand importance and should be considered, such as whether the article was published on the home page or not. According to this logic, if source A is 10 times more important than source B, prevalence of a brand mentioned by A will be 10 times higher than prevalence of a brand mentioned by B. Weights of network links are also determined considering source weights, and filtered accordingly.

Similarly, the analyst might want to limit the analysis to the initial part of online news, considering that most readers stop before reading 30% of webpages [24] and that a brand that appears in the title of an article is presumably more relevant than one only appearing at the end of its body. The SBS BI app offers the possibility of limiting the portion of text that will be analyzed.

Brand Intelligence Dashboard

Some of the most relevant information obtained from the analysis is summarized by a graphical dashboard. Some of the main graphs included in the dashboard are described in the following section, whereas examples are provided in Sect. 4. All graphs are interactive and were created via the Plotly[5] library, excluding the topic network which has been generated using Cytoscape Js [25]. The app was mainly programmed using the Python language.

SBS Time Trends

The SBS Time Trends interactive line graph shows the dynamic evolution of the Semantic Brand Score for each brand over time. In a second tab, absolute values are replaced by proportional values with respect to competitors (see Fig. 10.2).

Brand Positioning

This is a scatter plot with the SBS on the vertical axis and brand sentiment on the horizontal axis (Fig. 10.3). Combining information from these two measures we can

---

[4]https://www.alexa.com/siteinfo.

[5]https://github.com/plotly/plotly.py.

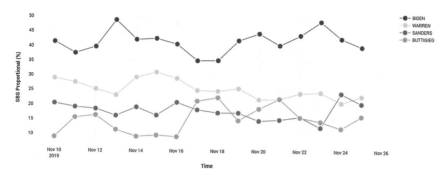

Fig. 10.2   SBS proportional time trends

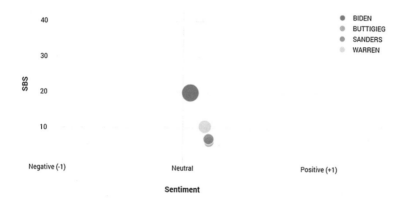

Fig. 10.3   Brand positioning

have an idea of brand positioning, with the most important brands being located in
the top right part of the graph (high importance and positive sentiment). Different
approaches are possible for the calculation of sentiment. The app default for the
English language is to use the VADER lexicon included in the NLTK library [26].
Sentiment varies between -1 and $+1$, where -1 is negative and $+1$ is positive.
It is important to notice that SBS BI calculates sentiment considering the sentences
related to each brand, and not average scores of full documents. This is important, for
example, to make a distinction in the case of a text where two brands are mentioned
and the author is speaking in a positive way about a brand and negatively about the
other. Punctuation is considered in the calculation of sentiment.

Average SBS Scores

The third graph offers a visualization of the average SBS scores obtained considering
the full time of analysis (Fig. 10.4). It is a stacked bar chart which shows the contri-

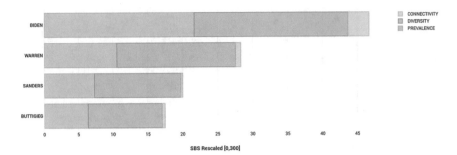

Fig. 10.4 Average SBS scores

bution of prevalence, diversity and connectivity to the final score. Each measure is rescaled in the interval [0, 100]. A more appropriate evaluation of the overall importance of each brand could also be obtained repeating the analysis on a single time interval, including all text documents available.

Most Common Words and Brand Associations

It can be interesting for the analyst to know the most frequent words used in a specific timeframe or overall, to discover concepts, people and events that were typically mentioned in a text corpus. SBS BI provides this information, through a dynamic sunburst graph. In addition, other charts show the top textual associations with the analyzed brands. Looking at the most frequent word co-occurrences, the user can understand the textual image of the brand, and its related message (Fig. 10.5). In order to identify the main traits that distinguish a brand from competitors, the app also shows unique associations.

Brand Image Similarity

In this chart, the more similar is the textual image of two brands, the closer they appear (Fig. 10.6). The user can get an overall view of the similarity of the words that co-occur with different brands. This can be seen as a proxy of the brand image of text authors, or could be used to assess similarity of communication strategies, if texts are authored by companies. Cosine similarity is the metric used [27], together with multidimensional scaling [28], in order to plot the graph in two dimensions.

Target Words for Connectivity and SBS Improvement

In addition to measuring the importance of a brand, it is also useful to understand what actions can be taken in order to improve this score. Prevalence increases if a brand is

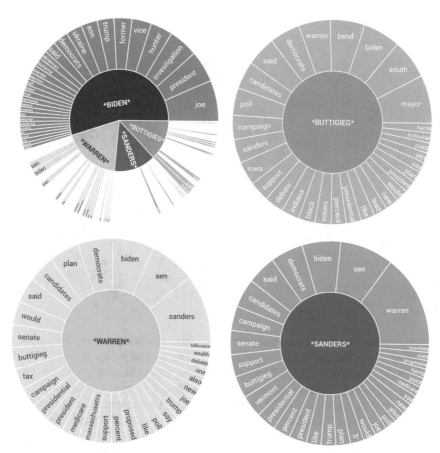

Fig. 10.5 Textual brand associations

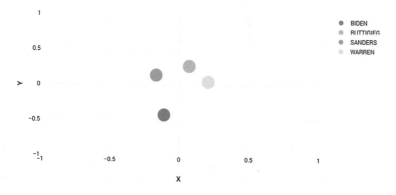

Fig. 10.6 Image similarity

Fig. 10.7 Target words

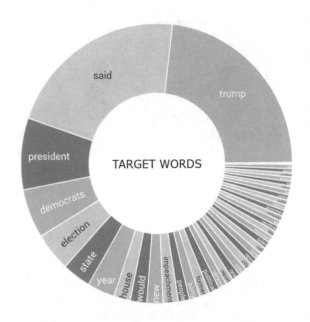

frequently mentioned. Accordingly, the press office of a company, or the campaign office of a political candidate, could work to obtain more media coverage. When diversity is low, its value can be increased by linking the brand name to heterogeneous themes and concepts. However, designing a strategy to improve connectivity, that is, the brand "brokerage power", is less easy. Brand managers need to find those words that, if used in future communication, could potentially make their brand more central in the discourse. However, they should also avoid favoring competitors and pay attention to keeping communication consistent with their brand strategy. In terms of graph theory, this is a maximum betweenness improvement problem [29], with additional constraints—such as the presence of forbidden nodes and opponents. Specific algorithms are implemented in the SBS BI app to solve this problem. The best set of words is shown by the target words graph (Fig. 10.7) and it can be customized for each brand. These are the words that, if connected to a brand, have the highest potential to increase its connectivity.

Main Discourse Topics

Topic modeling is a popular theme in text mining [17], with some of the most common approaches using Latent Dirichlet Allocation [30]. The goal is to automatically extract the main discourse topics from a set of documents and represent them through their most salient words. The SBS BI app reaches this goal using a different methodology, that is, through the clustering of the full co-occurrence network. After the removal of isolates and negligible links, the Louvain algorithm [31] is used to determine the main network clusters (other approaches are also possible). Words that better represent each cluster are subsequently identified through the following formula:

$$IW_i^K = \sum_{\substack{j \in K \\ j \neq i}} w_{ij} \frac{\sum_{\substack{j \in K \\ j \neq i}} w_{ij}}{\sum_{\substack{j = 1 \\ j \neq i}}^{N} w_{ij}} = \frac{\left(\sum_{\substack{j \in K \\ j \neq i}} w_{ij} \right)^2}{\sum_{\substack{j = 1 \\ j \neq i}}^{N} w_{ij}}$$

where IW_i^k is the importance of the word i belonging to the cluster K, N is the number of nodes in the network, and w_{ij} is the weight of the arc connecting nodes i and j (other approaches are also possible). The idea is that the most representative words are those with many strong connections within the cluster and a low proportion of links to nodes outside the cluster—similar to the logic of modularity functions [32]. The topic modeling graph is presented in Fig. 10.8. This graph also helps identify which topic is closest to each brand (red nodes) and the strength of connection between the topics. The app also calculates the importance of each topic and the weight of its connections to the different brands.

Lastly, some other charts are produced by the app, such as the time trend of the number of unique brand associations. The app allows the analyst to generate new, customized reports using the results files, which can be downloaded at the end of the analysis.

Case Study: 2020 US Democratic Primaries

The field of 2020 Democratic presidential candidates has been defined by many commentators as the largest Democratic primaries field in modern history, since it involved more than two dozen candidates and included six female candidates. As of November 24, 2019, a total of 18 candidates were seeking the Democratic presidential nomination in 2020.

On November 26, 2019 we downloaded 50,000 articles from the Event Registry database [20]–which contains both mainstream and blog news collected from around the world. We selected the most recent articles which were related to the 2020 Presidential Race and the Democratic Primaries. Articles were published in the USA in the period November 10–25, 2019. Using the SBS BI app, we generated reports for the top four candidates that had a vote share higher than 5% in the last available national polling average [33]. These candidates were Joseph R. (Joe) Biden Jr., Elizabeth Warren, Bernie Sanders and Pete Buttigieg.

Figure 10.2 illustrates the time trends interactive graph for the selected candidates, with fluctuating dynamic positions over time. We notice that Biden's positioning is

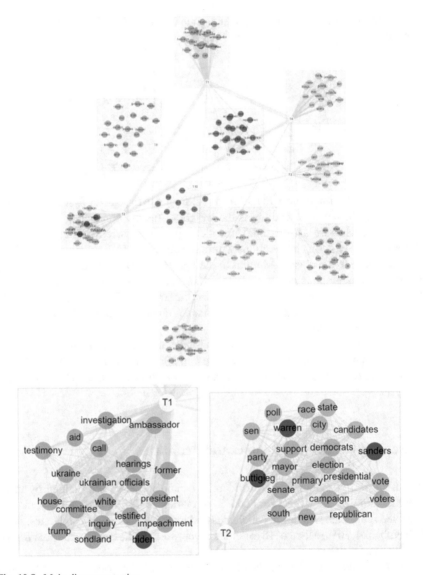

Fig. 10.8 Main discourse topics

constantly higher than the others, which indicates a higher frequency with which the Biden name appears in the online news, but also a higher diversity and connectivity. This can be explained by the events associated with Biden's son in Ukraine, the subsequent impeachment process for President Trump, in addition to the internal discussion with the rest of the primaries candidates. Conversely, the SBS trends for the others are lower, though with a higher fluctuation. In particular, SBS trends for

Buttigieg and Sanders are more intertwined, which might indicate that online news report stories about them that are highly associated.

The scatter plot in Fig. 10.3 combines information from the SBS values (y-axis) and the brand sentiment values (x-axis). The online discourse around Biden is the most different in terms of variety of news reported about him. Warren is the second most reported candidate with Buttigieg and Sanders immediately after. In terms of sentiment analysis, the online news present the front-runner (Biden) in a more neutral way, perhaps due to the Trump-Ukraine scandal and his son's involvement overseas. While Biden is associated with a more diverse set of topics (both positive and negative), the other three candidates are associated with more positive words.

Figure 10.4 shows the stacked bar chart with the average SBS scores during the entire time interval (November 10–25). We notice that Sanders has the lowest relative value of connectivity, which means that the brand "Sanders" serves fewer times as an indirect link between all the other pairs of words in the co-occurrence network. In other terms, the Sanders brand does not support an indirect connection between political concepts that are not directly co-occurring. This could indicate that Sanders was reported on via online media as talking about a specific set of agenda points, without much connection to other clusters of concepts. On the contrary, the Biden brand has the highest indirect link between all the other pairs of words, as it appears on the media as connecting clusters of concepts that are not directly connected to each other.

The textual brand associations, which are illustrated in Fig. 10.5, present some interesting insights for the selected candidates. If we zoom into the discussion around Joe Biden, we notice some of the most frequently used words in the specific time-frame: "Burisma", "Hunter", "investigation" and "son" were all more frequently mentioned with the candidate "Biden" in the online news. It is not a surprise that these were the top textual associations with the "Biden-brand", since Burisma is the holding company for a group of energy exploration and production companies based in Ukraine where Biden's son, Hunter, worked and was a board member. The topics associated with the other candidates were more diverse and refer directly to the specific agenda points that the candidate would bring to the table in a future presidential race. For example, the textual brand association of Elizabeth Warren highlights largely discussed points such as "Medicare, wealth, and billionaire", which are indeed the key differentiators of the Massachusetts senator. The textual association for Sanders was somewhat similar to Warren's, as Sanders co-occurred with concepts such as "progressive and Medicare".

If we look at the unique associations of words to each brand/candidate, it is interesting how Buttigieg has a strong association with "combat, nonwhite, qualified and gay". This is not surprising since Buttigieg–former Mayor of South Bend, Indiana, and a veteran of the War in Afghanistan–would be the first openly gay president, if elected, and has received strong support from the "non-white" part of the Democratic base. Former President Barack Obama once called him the "future of the Democratic Party".

The report also offers some insights in terms of image similarity. The graph in Fig. 10.6 indicates that in terms of image similarity Biden differentiates himself in a more significant way, likely because of the candidate's association with the Ukrainian investigation involving the current President, Donald Trump. Warren, Buttigieg and Sanders tend to be reported more often as sharing a similar political discourse. This is an important insight for their political campaign since voters might not be able to distinguish one candidate from the other unless they become more specific with their positions.

Sanders, Buttigieg and Warrens' positions on mental health and health policy show how their images are similar to each other and far from Biden. While the candidates have focused their dominant themes on universal health care, climate change and reproductive rights, Vice President Joe Biden has been slower to embrace marijuana law reform and the legalization of cannabis for medical purposes [34], which could help veterans avoid or alleviate substance abuse disorder.

The target words graph in Fig. 10.7 illustrates how the online discourse of these candidates is predominantly focused on the current President, Mr Trump. The online press is frequently reporting the direct statements of the politicians (see the highly reported word "said"), as well as topics such as "impeachment, Ukraine, State, House (of Representatives)". These concepts act as catalyst of connectivity between the candidates and the rest of the political discourse.

Figure 10.8 shows the topic modeling graph. The network of keywords indicates what the main discourse topics are and which is closer to each brand. The brighter red nodes are the brand/candidates. Very interestingly, Biden is mainly embedded in a cloud of words (T1) that is separate from the clouds where Warren, Buttigieg and Sanders are reported (T2). The width of the link connecting T1 and T2 is big enough to suggest that all four candidates are associated to and talking about a sub-set of shared concepts. However, the nature of the words associated with Joe Biden shows a strong dissimilarity from the ones associated with the other candidates. Biden is reported in articles associated with "testify, impeachment, hearings, ambassador", which clearly refer to the Trump-Ukraine scandal. The cluster in which the other candidates are mainly embedded into is characterized by words associated with the primaries vote and election.

Another interesting insight from the topic modeling network is the extremely weak connection between the four candidates' clusters (T1 and T2) and the cluster T6, reporting topics such as "Israel strikes in Gaza, rockets, police action in Hong Kong, political unrest in Bolivia". This seems to suggest that all four candidates are currently focusing on national policies rather than foreign policy issues. We would expect that, at a later stage of the primaries process, the emerging front-runners will be asked their opinion on foreign policy issues, which are important if they plan to become the next President of the United States.

Table 1 offers a comparison of brand importance in online news and poll results. The first column represents the percentages for each candidate based on the most recent poll. In the second column we reported the same values calculated as if these four candidates were the only ones to run. The third column reports the values of the predictions using SBS dimensions based on the online news sources [1]. The last column is the most interesting as it illustrates the difference between the values of the adjusted polls and the proportional SBS. This seems to indicate that the online

Table 10.1 Comparing brand importance in online news and poll results

Candidate	Average last polls (%)	Adjusted polls (%)	Proportional SBS (%)	Difference (prop. SBS—Adj. Polls) (%)
Biden Jr.	27	36.0	40.0	4.0
Warren	22	29.3	23.1	−6.2
Sanders	18	24.0	20.6	−3.4
Buttigieg	8	10.7	16.3	5.6

discourse around these four candidates is translating into different proportional and relative impacts on their polls. While Pete Buttigieg is driving relatively fewer voters based on the polls, the media is reporting him as relatively more important and connected to a variety of topics.

Discussion and Conclusions

In this chapter we have presented new methods to measure and assess the importance and relative positioning of brands. To explain the functionalities and reports available via the SBS BI app, we have discussed how the four front-runners for the US Democratic 2020 primaries are positioned in the online news.

The SBS BI app represents an innovative tool to measure brand importance and brand positioning, combining the components of the SBS indicator (prevalence, diversity and connectivity) and relying on methods and tools of text mining, sentiment analysis and social network analysis.

Overall, the application of the SBS BI to a limited time period of the US democratic primaries indicates that Joe Biden is the one with the richer textual embedding, spanning boundaries of political discourse. The method we describe in this paper has the potential to complement traditional polls, by providing a comprehensive analysis of what people (news reporters, but also commentators, voters etc.) say about the candidates online. Our method is based on the automatic mining of big (textual) data, which could help counteract the so-called "pollster fatigue", where voters start to avoid answering the calls of pollsters, impacting the representativeness of the sample.

The SBS BI app—not limited to the analysis of political news—is in continuous development and we plan to add more functionalities in the near future. For example, we plan to improve the algorithm used for the identification of target words, to enrich the set of recommendations the app can provide to increase brand importance. Topic modeling through the clustering of co-occurrence networks still has open research questions, as well as the identification of the most salient words for each topic.

References

1. A. Fronzetti Colladon, Forecasting election results by studying brand importance in online news. Int. J. Forecast. **36**, 414–427 (2020). https://doi.org/10.1016/j.ijforecast.2019.05.013
2. A. Fronzetti Colladon, The semantic brand score. J. Bus. Res. **88**, 150–160 (2018). https://doi.org/10.1016/j.jbusres.2018.03.026
3. A. Fronzetti Colladon, F. Grippa, R. Innarella, Studying the association of online brand importance with museum visitors: An application of the semantic brand score. Tour. Manag. Perspect. **33**, 100588 (2020). https://doi.org/10.1016/j.tmp.2019.100588
4. S.L. Warner, Randomized response: a survey technique for eliminating evasive answer bias. J. Am. Stat. Assoc. **60**, 63–69 (1965)
5. K. Olson, Survey participation, nonresponse bias, measurement error bias, and total bias. Public Opin. Q. **70**, 737–758 (2006)
6. U. Grandcolas, R. Rettie, K. Marusenko, Web survey bias: sample or mode effect? J. Mark. Manag. **19**, 541–561 (2003)
7. P.A. Gloor, J. Krauss, S. Nann, K. Fischbach, D. Schoder D, Web science 2.0: identifying trends through semantic social network analysis, in *2009 International Conference on Computational Science and Engineering* (IEEE, Vancouver, Canada, 2009), pp. 215–222
8. P.A. Gloor, Coolhunting for trends on the web, in *2007 International Symposium on Collaborative Technologies and Systems* (IEEE, Orlando, FL, USA, 2007), pp. 1–8
9. D. Godes, D. Mayzlin, Using Online conversations to study word-of-mouth communication. Mark. Sci. **23**, 545–560 (2004)
10. K.L. Keller, Conceptualizing, measuring, and managing customer-based brand equity. J. Mark. **57**, 1–22 (1993)
11. T.H. Bijmolt, M. Wedel, R.G. Pieters, W.S. DeSarbo, Judgments of brand similarity. Int. J. Res. Mark. **15**, 249–268 (1998)
12. D.J. Langley, N. Pals, J.R. Ortt, Adoption of behaviour: predicting success for major innovations. Eur. J. Innov. Manag. **8**, 56–78 (2005)
13. P. Marsden, Brand positioning: meme's the word. Mark. Intell. Plan. **20**, 307–312 (2002)
14. M.M. Mostafa, More than words: social networks' text mining for consumer brand sentiments. Expert Syst. Appl. **40**, 4241–4251 (2013)
15. D.A. Aaker, Measuring brand equity across products and markets. Calif. Manag. Rev. **38**, 102–120 (1996)
16. L.G. Cooper, A review of multidimensional scaling in marketing research. Appl. Psychol. Meas. **7**, 427–450 (1983)
17. D. Blei, Probabilistic topic models. Commun. ACM **55**, 77–84 (2012)
18. A.N. Srivastava, M. Sahami (eds.), *Text Mining Classification, Clustering, and Applications* (Chapman & Hall/CRC, Boca Raton, FL, 2009)
19. O. Netzer, R. Feldman, J. Goldenberg, M. Fresko, Mine your own business: market-structure surveillance through text mining. Mark. Sci. **31**, 521–543 (2012)
20. G. Leban, B. Frtuna, J. Brank, M. Grobelnik, Event registry—learning about world events from news, in *Proceedings of the 23rd International Conference on World Wide Web* (2014), pp 107–110
21. S. Bird, E. Klein, E. Loper, *Natural Language Processing with Python* (O' Reilly Media, Sebastopol, CA, USA, 2009)
22. W. De Nooy, A. Mrvar, V. Batagelj, *Exploratory Social Network Analysis with Pajek*, 2nd edn. (Cambridge University Press, Cambridge, MA, 2012)
23. A. Fronzetti Colladon, M. Naldi, Distinctiveness centrality in social networks. PLOS ONE **15**, e0233276 (2020). https://doi.org/10.1371/journal.pone.0233276
24. J. Nielsen, How little do users read? in Nielsen Norman Group (2019). https://www.nngroup.com/articles/how-little-do-users-read/
25. M. Franz, C.T. Lopes, G. Huck, Y. Dong, O. Sumer, G.D. Bader, Cytoscape.js: a graph theory library for visualisation and analysis. Bioinformatics **32**, 309–311 (2016)

26. C.J. Hutto, E. Gilbert, VADER: a parsimonious rule-based model for sentiment analysis of social media text, in *Proceedings of the Eighth International AAAI Conference on Weblogs and Social Media* (AAAI Press, Ann Arbor, Michigan, USA, 2014), pp. 216–225

27. A. Huang, Similarity measures for text document clustering, in *Proceedings of the sixth New Zealand computer science research student conference (NZCSRSC2008)* (Christchurch, New Zealand, 2008), pp. 49–56

28. A. Mead, Review of the development of multidimensional scaling methods. Statistician **41**, 27 (1992)

29. G. D'Angelo, L. Severini, Y. Velaj, On the maximum betweenness improvement problem. Electron. Notes Theor. Comput. Sci. **322**, 153–168 (2016)

30. D. Blei, A. Ng, M. Jordan, Latent dirichlet allocation. J. Mach. Learn. Res. **3**, 993–1022 (2003)

31. P. De Meo, E. Ferrara, G. Fiumara, A. Provetti, Generalized Louvain method for community detection in large networks, in *2011 11th International Conference on Intelligent Systems Design and Applications (ISDA)*, ed. by S. Ventura, A. Abraham, K. Cios, C. Romero, F. Marcelloni, J.M. Benítez, E. Gibaja (IEEE, Córdoba, Spain, 2011), pp. 88–93

32. U. Brandes, D. Delling, M. Gaertler, R. Gorke, M. Hoefer, Z. Nikoloski, D. Wagner, On modularity clustering. IEEE Trans. Knowl. Data Eng. **20**, 172–188 (2008)

33. J.C. Lee, A. Daniel, R. Lieberman, B. Migliozzi, A. Burns, Which democrats are leading the 2020 presidential race? The New York Times (2019)

34. T. Angell, Sanders, Warren, Biden and Buttigieg Include Medical Marijuana, in Veterans Day Plans. Forbes (2019)

Chapter 11
Finding Patterns Between Religions and Emotions

A Quantitative Analysis Based on Twitter Data

Sonja Fischer, Alexandra Manger, Annika Lurz, and Jens Fehlner

Abstract The emotions someone associates with his or her religion and how this person talks about his or her faith have always been considered a personal topic. In this paper, the question of whether specific religions and emotions are connected is discussed. Based on Twitter data, individual networks, or so-called "tribes", are created for four religions: Buddhism, Christianity, Islam and Judaism and four emotions: anger, fear, joy and sadness. Similarities and differences between tribes are analyzed using the content of the tweets. A network analysis is done for all tribes and the resulting data is used to create a machine learning model for each category. Using these, general patterns between emotions and religions are outlined and discussed. An analysis with further data was conducted on our model.

Introduction

Religion has always played an important role in people's lives. At the beginning of mankind, everything unexplainable was usually attributed to one or more Gods [5]. In the Middle Ages, when the church was one of the two great powers, people had to have a faith. People without any religion or with beliefs that were different were often punished and killed. However, faith was never a pure coercion. Most people believed in their respective Gods out of conviction [7]. Today, though freedom of religion exits in most places in the world, it is not surprising that there are still many people who believe in God [11].

Unfortunately, despite this freedom, discrimination and hatred toward certain religions have risen considerably. In the Western world, the popularity of conservative and anti-immigration ideologies has been growing steadily, which on the other hand has resulted in discrimination against certain religions. Despite the rise in religious discrimination, people have held onto their faith. Possible reasons for this may include beliefs that faith gives them strength and guidance through their lives, fear

S. Fischer (✉) · A. Manger · A. Lurz · J. Fehlner
Otto-Friedrich-Universität Bamberg, Bamberg, Germany
e-mail: sonja1.fischer@stud.uni-bamberg.de

© Springer Nature Switzerland AG 2020
A. Przegalinska et al. (eds.), *Digital Transformation of Collaboration*,
Springer Proceedings in Complexity,
https://doi.org/10.1007/978-3-030-48993-9_11

of death and consequences of non-eternal life, and adherence to family traditions. In a scientific environment, it would be interesting to explore the emotions of specific religious groups and not just how other groups feel about them. Inversely, it would also be valuable to know if a specific emotion has a connection to a faith.

Related Work

The connection between religions and emotions has already been considered in substantial research. Mostly, however, it is discussed in the literary field or addressed by surveys and observations. In addition, the focus lies often more on spiritual people than on religious ones.

In the chapter "Religion, emotions, and health" of the book "Handbook of Emotion, Adult Development, and Aging" by McFadden and Levin, the authors argue that religion can cause positive emotions that bring salvation from a psychological perspective [8].

"The Oxford Handbook of Religion and Emotion" discusses both: the mutual influence of religions and emotions, the culture of religions and that religious culture gives rise to emotions [2]. The introduction "The Study of Religion and Emotion" analyzes the emotional component in religion. It is emphasized that religions strongly influence emotions and therefore, religion affects behaviors through emotions. In comparison to many previous studies in this field, modern results have been taken into consideration. There is some progress in the field of emotion research which simplifies to categorize them. In addition, this study already distinguishes between different religions, as well as gender, age and other unambiguous characteristics. It also addresses specific emotions such as love, hope, ecstasy, melancholy and terror in more detail. While this study concludes that feelings and religious affiliation correlate, the practical component is missing from the research [1].

Another suitable study is "Positive emotions as leading to religion and spirituality" written by Saroglou, Buxant and Tilquin. In this study, three groups "Without Faith", "Religious" and "Spiritual" were distinguished. The results show that both religious and spiritual subjects had more positive feelings during a conversation, although the result was clearer for spiritual subjects. The study suspected that positive emotion reinforced faith. In this study, however, fewer than 200 people were examined. In addition, no distinction was made between the different religions [12].

The most current study found is from 2014 and was written by Van Cappellen, Toth-Gauthier, Saroglou and Fredrickson. In the empirical study, the positive emotions of religious, spiritual and non-religious people were compared. Two groups were examined. The first group consisted of subjects from European churches and the second consisted of American workers from a university who were interested in meditation. The results show that religious people have more feelings of reverence, gratitude, love and peace, and less feelings of pleasure and pride. However, subjects were directly investigated in the study and the results based on their self-perception. In addition, the study made no distinctions between different religions [14].

Our work examines the mutual relationship between four religions and four emotions in detail. One positive as well as three negative emotions are considered. Since this paper is based on Twitter data, the self-perception bias is limited. The examination of tweets allows access to a lot of profiles and their data. Based on this, our research question is: *Are there patterns between specific religions and emotions?*

In order to answer this question, we created eight tribes with relevant Twitter accounts: four different religion tribes and four different emotion tribes. Based on the datasets, we could analyze the tribe network. The generated data was used in a machine learning approach to find patterns between religions and emotions.

Methods

An explorative approach was chosen for this project. The NRC-Affect-Intensity-Lexicon was used to create the emotion tribes [9]. The religion tribes were created by searching famous leaders for each religion and their followers. Network measurements were calculated based on the six honest signals of collaboration [4].

Differences between the word usage are shown by a content analysis of the tweets. We created two Machine Learning models: one for the religion tribes and one for the emotion tribes. The approach and tools that were used in the different project iterations will be discussed further in the following sections.

Tribe Creation with Galaxy Scope

The website galaxyscope.galaxyadvisors.com provides the Tribe Creator which was used for the tribe creation [3]. Relevant English-speaking accounts were found on Twitter and added via the Twitter Profile Search to a dataset, so-called tribes. Four religion tribes and four emotion tribes were created: Anger (156 members), Buddhism (160 members), Fear (155 members), Christianity (241 members), Joy (264 members), Islam (178 members), Sadness (154 members) and Judaism (172 members). These four emotions were chosen because of the following two reasons. At first machine learning models are more precise if there are less categories. The second reason is the difficulty to distinguish between profiles of specific emotions, e.g. the tribes of Anger and Fear have many common words. The more emotions we include, the harder is the distinction of those emotions.

The Twitter accounts were manually analyzed and checked to verify that only suitable Twitter accounts were part of a tribe. The Tribe Creator also creates Hashtag clouds that contain the most frequently used words for a tribe. These were used for validating the reasonableness of the tribe members.

For the religion tribes, well-known personalities (e.g. most famous rabbis or Buddhist monks that are listed in Wikipedia), obvious keywords or keyword phrases (e.g. I am a convinced Christian), and unambiguous hashtags (e.g. #jesusislove) were searched. The emotions were chosen based on the NRC-Affect-Intensity-Lexicon [9]. The words associated with four emotions: anger, fear, joy and sadness were used to find new tribe members.

Analyzing Data in Condor

We used the software Condor to analyze and visualize our tribe networks. With the "Fetch Tribes" function of Condor, it was possible to easily import our Tribe Creator data. Furthermore, Condor added further members to our manually generated tribes. We processed the datasets and calculated network measurements: Centrality annotations, Betweenness centrality, Degree centrality, Contribution index oscillation, Contribution index annotations, TurnTaking annotations, Calculate sentiment, Calculate influence, Pennebaker Pronoun Frequency, Tribefinder Annotation (necessary to check if and how our tribes are categorized into specific "Alternative Realities"). Condor provides network visualization and word clouds (positive sentiment: green; negative sentiment: red). Based on these, we checked our tribes for errors.

Machine Learning with Rapid Miner

Using RapidMiner, the resulting datasets could easily be processed. We first selected 24 attributes for the machine learning process: Centralities (Betweenness centrality and Degree centrality), Tribefinder Annotations (Personality, Lifestyle, Alternative-Realities, Recreation and Ideology), Frequency-Attributes (was, my, it, in, the, with, to, but, for, have, and, me, you), Complexity, Sentiment, Contribution index and Emotionality. We trained a random forest model and reached an accuracy of 83.49% for religions (Table 11.1) and 87.07% for emotions (Table 11.2). The split validation training method was used.

The accuracy of the random forest model was highly dependent on attribute selection. Based on our data, the accuracy would rise to 100% for religions and 99.99%

Table 11.1 Accuracy for religion model: 83.49%

	True Buddhism	True Christianity	True Islam	True Judaism	Class precision (%)
Pred. Buddhism	18461	943	677	11	91.88
Pred. Christianity	0	13733	0	0	100.00
Pred. Islam	1519	5679	19857	3999	63.9
Pred. Judaism	554	188	0	16524	95.70
Class recall	89.90%	66.88%	96.70%	80.47%	

Table 11.2 Accuracy for emotion model: 87.07%

	True anger	True joy	True fear	True sadness	Class precision (%)
Pred. Anger	21165	1199	3080	225	82.45
Pred. Joy	819	16694	66	449	92.60
Pred. Fear	0	0	18687	0	100.0
Pred. Sadness	518	4609	669	21828	79.02
Class recall	94.06%	74.19%	83.05%	97.00%	

Table 11.3 Results of additional testing (Buddhism classification)

Index	Nominal value	Absolute count	Fraction
1	Buddhism	41156	0.747
2	Judaism	5455	0.099
3	Islam	4302	0.078
4	Christianity	4150	0.075

for emotions if all attributes we analyzed in Condor were included in the machine learning process.

The cross-validation method was applied during the training. It did not show significant differences in accuracy ($\pm0.75\%$ for the religions and $\pm1.14\%$ for the emotions) and the model trained with the split validation was later used for the classifications.

After generating the random forest model, it could also be applied to data that had not previously been trained on in order to get a better understanding of the fit of the model. Table 11.3 shows that the classification for Buddhism worked well with a fraction of 0.747 classified to the correct tribe.

The summary of all tribes' fractions that were classified correctly is as follows: Anger (0.623), Buddhism (0.747), Fear (0.706), Christianity (0.605), Joy (0.468), Islam (0.767), Sadness (0.967) and Judaism (0.392). With the help of the emotion random forest model and religion random forest model, the emotions and religions of different samples could now be predicted.

The religion model was applied to the tweets of the emotion tribes in order to classify their religion. This was also done in reverse order (classification of emotions for religious tribes). Based on this we identified patterns between religions and emotions.

Word Usage Analysis and Visualization with RStudio

In order to get an overview of which tribes were similar, the words used were compared tribe-wise. The word frequency percentage was calculated for each tribe by dividing the occurrence of a word in a tribe by the occurrence in both tribes. These percentages were then displayed using a scatterplot. Each point signifies an individual word, the size how often the word was used in total and the location of the percentage of usage in both tribes. Furthermore, R word clouds were created for a better display of word frequency in individual tribes.

Data Aggregation and Visualization of Machine Learning Results with RStudio

The results of the tweet classification are displayed in two separate stacked bar charts. Furthermore, an aggregation of the data was done. The sample was grouped by Twitter accounts. An account is classified based on the most frequent prediction of its tweets. This was also visualized in two stacked bar charts.

Results and Discussion

The preliminary findings, as well as a short discussion about them, are shown in the following paragraphs. Afterwards, the results of the prediction of religions for the emotion tribes and emotions for the religion tribes are presented. Furthermore, the result of applying the created models on Anti-Gun Control and Pro-Gun Control tribes as well as the Anti-LGBT and Pro-LGBT tribes are depicted.

Network Analysis

Most of the religion actors belonged to the Christianity tribe. A big part of the emotion network was represented by members with the emotion joy or sadness. We processed each dataset with already existing tribe datasets from Condor. The Alternative Reality tribes were of special interest for us because they include Spiritualism, Treehugger, Nerd and Fatherlander. All of our four religion tribes fit perfectly into the existing Spiritualism tribe. The same analysis was done with the emotion tribes, but here only Sadness belongs to Spiritualism. The other emotions show a combination of all Alternative Reality tribes.

Word Usage

The word clouds show several prominent points. For example, in the Buddhism tribe, it is apparent that the words "Buddhism", "meditation", "Buddhist" and "mind" are words often tweeted. In comparison to the other religions, the usage of words that are connected to daily Buddhist practices stands out in this tribe. The words "meditation", "mind", "dharma", "practice" and "compassion" are all related to the Buddhist lifestyle. Christians and Muslims use their words for god ("god"/"Allah", "god", "lord") and for prophet ("Jesus", "Christ", "lord"/"prophet", "Muhammad"). The Judaism tribe tweets more about itself as a community and country ("Israel"), not as a faith and its principles and components. The usage of Yiddish and Hebrew words is also apparent. Overall, it can be seen that the content of the tweets fits the category they have been assigned. We can see that the tribes created are correct and the data can be used for further analysis.

For the emotion tribes, the Joy tribe has words that signify joy like "love", "happy", "beautiful" and "friend" and therefore fulfills its desired use. The same is observable in the Sadness tribe that tweets about "depression", "stillbirth", "miscarriages" and "mental". However, a point of criticism can be that words like "hope", "support" and "love" also occur. This can be attributed to the fact that these are tweets of specific accounts over a long period of time. Individuals are likely to have emotions other than sadness. Another reason is that support groups are also part of this tribe. Their tweets give advice to counteract sadness, not just talk about it.

The Fear and Anger tribes seem relatively similar in the words they used. These tribes mostly used words like "realdonaldtrump", "president" and "trump". However, the Fear tribe also talks about "mentalhealth", "war" and "anxiety". Furthermore, the Anger tribe additionally has words like "shit", "bad" and "hate". The two tribes are dominated by political topics during the time when the snapshot was taken (around the congress election in the US in 2018). The policies of the American president and his administration's conduct created a lot of anger between his supporters and critics. This can be observed in the word usage of the Anger tribe. The political topics in the Fear tribe can also be connected to this behavior. The discrepancy of the president's demeanor in comparison to previous presidents and his tendency to announce policies in tweets seem to cause fear for some Americans, which is why political topics are a part of this tribe.

To validate the emotion tribes, additional word clouds were created in Condor. This approach offered the opportunity to identify whether the words were used in a positive or negative context. The wordings of fear and anger were similar and further reflected similar sentiments: trump (positive context), people (negative context), democrats (negative context), president (positive context). In the joy cloud almost all words except "southpark22" stood in a positive context. "southpark22" was probably used in context with sarcasm and/or insults and therefore is marked red. Sadness is a combination of both, words in a positive and negative context. Overall, it should be mentioned that the tweets selected were based on the actors in the tribe. Even though these actors experience more than one emotion, they show mainly the specific emotion. Correspondingly the emotion tribes mainly show the targeted emotion.

To further analyze the differences and similarities between tribes in one category and between emotions and religions, we did a pairwise comparison of tweeted words using R. The diagram shows two tribes, one on the y-axis and another one on the x-axis. The points in blue symbolize different words used in these tribes. The size of the dots shows the overall occurrence of a word and the location symbolizes the percentage of the frequency associated with a specific tribe. Note that the data is not normalized, which results in a shift of the points in the direction of the tribe with more tweets. Since for each possible comparison, both axes have at least one point situated on them, we conclude that there are words for each tribe which are exclusive to that tribe. This means that the tribes differ in their usage of vocabulary.

Emotions and Religions

Figure 11.1 shows that a large percentage of Christians was predicted to experience mostly the emotion of anger. We assume that the result was caused by a high number of Christian accounts on Twitter in general and because the platform only supports short messages that seem to be often anonymous and impulsive.

Sadness is dominated by Muslims and Buddhists. Based on the comparison with the Alternative Reality "Spiritualism", Sadness and Spiritualism are connected with each other. We hypothesize that Buddhists and Muslims are more spiritual than the other religions (see 4.1).

Figure 11.2 shows the emotions predicted for religions. Overall, there is lots of sadness in all religion tribes. We suggest that sadness and negative events lead one to search for something bigger than oneself. Furthermore, there is no anger predicted for Buddhism at all. The largest proportion of fear is predicted for the Judaism tribe.

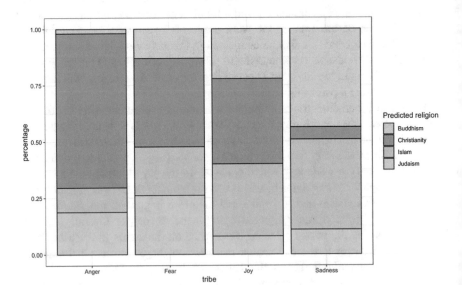

Fig. 11.1 Religions predicted for emotions in actor view

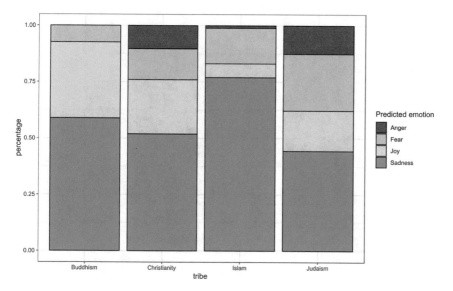

Fig. 11.2 Emotions predicted for religions in actor view

Further Application

The next sections describe further applications of our model. Two different tribe themes are classified according to religion and emotion. In each case pro- and anti-attitude are considered.

Pro-Gun Control and Anti-Gun Control

The topic of gun control is very emotional and volatile. Thus, the tribes of Pro- and Anti-Gun Control were analyzed with our emotion model (Fig. 11.3). The results show some indication about which emotion is predominant. The Pro-Gun Control tribe contains mostly emotion of anger, whereas, in the Anti-Gun Control tribe, the emotion of fear dominates. However, in each of the tribes both emotions, anger and fear, are presented with a high percentage. This could be due to the fact that the Pro-Gun Control group considers guns to be dangerous and evil without any benefits whereas the Anti-Gun Control group feels a threat to their safety and Second Amendment right to bear arms to protect themselves.

Figure 11.4 shows that for the Pro-Gun Control as well as for the Anti-Gun Control Tribe, the largest number of actors are Christians. There is also a conspicuous number of Jews predicted for Pro- and Anti-Gun Control. Buddhists are not predicted for Anti-Gun Control at all. This result suggests again that Buddhists are more peaceful (see 4.3).

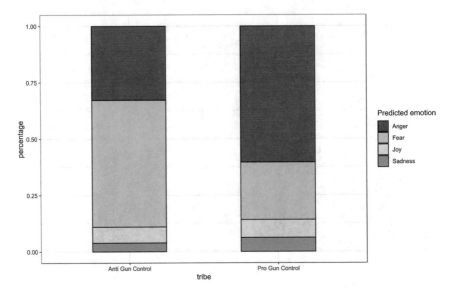

Fig. 11.3 Emotions predicted for gun control in actor view

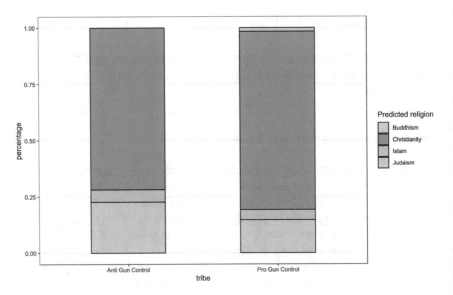

Fig. 11.4 Religions predicted for gun control in actor view

Pro-LGBT and Anti-LGBT

The emotion predominating Pro-LGBT is fear (Fig. 11.5). All other emotions are

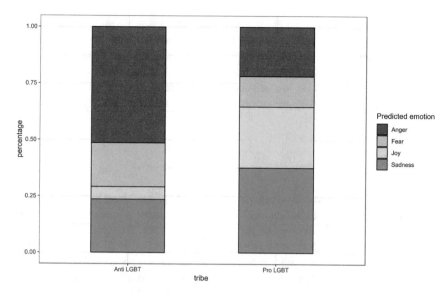

Fig. 11.5 Emotions predicted for LGBT in actor view

almost evenly distributed. Anti-LGBTs are predominated by anger and fear. The emotion joy is barely there.

The most predicted religion for the Anti-LGBTs and the Pro-LGBTs is Christianity (Fig. 11.6). We assume that, especially in the Western world, LGBT is a

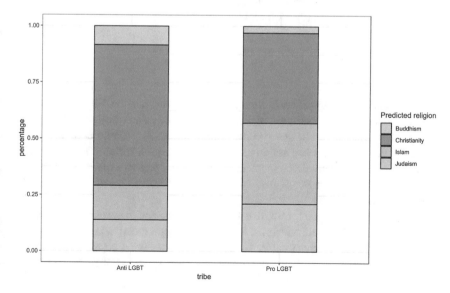

Fig. 11.6 Religions predicted for LGBT in actor view

controversial topic currently and therefore many Christians are involved. Christians are maybe less afraid of coming out and therefore, dominate the Pro-LGBT group. In both categories, there are almost no Buddhists predicted. For Buddhism, LGBT seems not to be a relevant topic. An interesting result is that Islam is predicted for about a quarter of LGBTs. We assume that especially modern Muslims have an account on Twitter, and they are proud of their open-minded ideology.

Conclusion

The aim of this paper was to discover conspicuous connections between different religions and emotions. Previous studies indicate that there is a connection between different religions and positive and negative emotions [6]. Therefore, our results of this explorative quantitative analysis based on Twitter networks give a first overview of the connections between specific religions and specific emotions. We also found a pattern between Spiritualism and Sadness in general which is mentioned in research [13]. Important findings in this context were highlighted and discussed. Hypotheses derived from our work and its results can be summarized as

H1a   *Sadness leads to Spiritualism.*
H1b   *Spiritualism leads to Sadness.*
H2a   *Sadness leads to Religion.*
H2b   *Religion leads to Sadness.*
H3   *Judaism is positively correlated to fear.*
H4a   *Buddhism is negatively correlated to anger.*
H4b   *Buddhism is positively correlated to joy.*

Limitations

Limitations of our work will be discussed in the following section. First, we assumed that tweets of one Twitter account include predominantly one emotion and therefore profiles were categorized in one specific emotion tribe. A different approach could be to analyze only tweets and assign a tweet to one emotion without regard to actors or Twitter profiles. Using this approach, the fact that a person experiences different emotions at different times and also tweets with different states of mind would be taken into consideration.

Our work only searched for English Twitter accounts and therefore other religions might be underrepresented. Muslims, for example, might mostly tweet in the Arabic language while Jews might tweet in Yiddish. English-speaking population is mainly Christian [10]. Besides the language limitation, some religions could be less active in using technologies than others. Buddhists may use platforms like Twitter less in general.

Although our dataset was very large, it is not ensured that our sample is representative of the chosen religions and emotions. Furthermore, the chosen religions and emotions could be extended (e.g. by Hinduism and surprise). By analyzing additional test data with RapidMiner (see 3.3), we realized that the classification of our machine learning model was not optimal. It ranged from a fraction of 0.392 (Judaism) to 0.967 (sadness).

Future Research

Our work represents a static analysis conducted at a specific point in time. Future research could analyze the tribes over a longer time period to check if certain events influence the emotion and religion tribes and how stable or easily influenceable they are.

Additional characteristics, e.g. impulsiveness or aggressiveness could be added to the emotion tribes to get more precise results. Additionally, an Atheism tribe could be added to see how Atheism differs from the religions regarding its emotions.

Our research is based on a network analysis and the tweet content is only indirectly included (e.g. represented by sentiment and emotionality) in the machine learning model. Further research could concentrate on language processing and use the tweet content as additional attributes for the model. It also would be interesting to check whether the machine learning algorithms would be more suitable for predictions with the used attributes.

Our research shows that members of the Sadness tribe are especially spirituals and every religion member belongs to Spirituality. Therefore, future research could focus on correlations of Spirituality and/or Religiousness and basic emotions. Additionally, correlations of sadness to Spirituality and/or Religiousness could be examined.

Additional qualitative research that interviews the examined persons behind the Twitter profiles would be interesting to experience how they interpret our findings. Doing this, emotions could be measured directly.

References

1. J. Corrigan, Introduction, in *The Study of Religion and Emotion,* ed. by J. Corrigan (Oxford Handbooks Online, 2009) pp. 3–13
2. J. Corrigan (ed.), *The Oxford Handbook of Religion and Emotion* (Oxford University Press, New York, 2009)
3. Galaxyadvisors: Galaxyscope (2018). https://galaxyscope.galaxyadvisors.com. Accessed 9 Feb 2019
4. P. Gloor, The signal layer: six honest signals of collaboration, in *Swarm Leadership and the Collective Mind* (2017), pp. 91–104
5. N. Jay, *Throughout Your Generations Forever: Sacrifice, Religion, and Paternity* (University of Chicago Press, 1992), p. 65

6. C. Kim-Prieto, E. Diener, Religion as a source of variation in the experience of positive and negative emotions. J. Posit. Psychol. **4**(6), 447–460 (2009)
7. F. Logan, *A History of the Church in the Middle Ages* (Routledge, Abingdon, Oxon, 2012)
8. S. McFadden, J. Levin, Religion, emotions, and health, in *Handbook of Emotion, Adult Development, and Aging*, ed. by, M. Carol, S.H. McFadden (1996), pp. 349–365
9. S. Mohammad, Word affect intensities, in *Proceedings of the 11th Edition of the Language Resources and Evaluation Conference (LREC-2018)* (Miyazaki, Japan, 2018)
10. C. Park, Religion and geography, in *Routledge Companion to the Study of Religion* Chapter 17 in ed by, J. Hinnels (London, Routledge, 2018), pp. 414–445
11. C. Park, *Sacred Worlds: An Introduction to Geography and Religion* (Routledge, London, 1994), p. 168
12. V. Saroglou, C. Buxant, J. Tilquin, Positive emotions as leading to religion and spirituality. J. Posit. Psychol. **3**(3), 165–173 (2008)
13. C.P. Scheitle, Bringing out the dead: gender and historical cycles of Spiritualism. Omega **50**(3), 237–253 (2004–2005)
14. P. Van Cappellen, M. Toth-Gauthier, V. Saroglou, B. Fredrickson, Religion and well-being: the mediating role of positive emotions. J. Happiness Stud. **17**(2), 485–505 (2014)

Chapter 12
Virtual Tribes: Analyzing Attitudes Toward the LGBT Movement by Applying Machine Learning on Twitter Data

Moritz Bittner, David Dettmar, Diego Morejon Jaramillo, and Maximilian Johannes Valta

Abstract In this paper, we investigate the application of machine learning techniques in the context of social media. Specifically, we aim at drawing conclusions from users' Twitter behavior and language to users' attitudes toward the LGBT movement. By using an adjusted procedure of the Cross Industry Standard Process for Data Mining (CRISP-DM) process, we create a prediction model for investigating and identifying those attitudes. Furthermore, we formulate step-by-step instructions for its deployment. We provide the reader with a theoretical background for our research domain and describe the methods that we use. Results show that there are two groups of contrary attitudes toward the LGBT community and that the language and behavior of users in the groups, respectively, differ from each other. Also, we identify word analyses as a valuable means for prediction. We also apply our model on another dataset to investigate its interspersion with the previously identified groups and demonstrate its effectiveness for predicting attitudes of a single actor on Twitter. Finally, we critically assess our findings and propose further fields of investigation in this area.

M. Bittner (✉) · D. Dettmar
Universität zu Köln, Albertus-Magnus-Platz, 50923 Cologne, Germany
e-mail: bittnerm@smail.uni-koeln.de

D. Dettmar
e-mail: ddettmar@smail.uni-koeln.de

D. Morejon Jaramillo · M. J. Valta
Universität Bamberg, Kapuzinerstraße 16, 96047 Bamberg, Germany
e-mail: diego-sebastian.jaramillo@stud.uni-bamberg.de

M. J. Valta
e-mail: maximilian-johannes.valta@stud.uni-bamberg.de

© Springer Nature Switzerland AG 2020
A. Przegalinska et al. (eds.), *Digital Transformation of Collaboration*,
Springer Proceedings in Complexity,
https://doi.org/10.1007/978-3-030-48993-9_12

Introduction

Since ancient times, *tribes* have been a popular concept in societies [5]. Tribes are groups of people that share the same language and values like culture and history. In particular, tribe members exalt their tribe above other tribes and groups, which leads to tribal consciousness and tribal loyalty (Cambridge Dictionary; Merriam-Webster Dictionary). The ancient tribes often lived among each other detached from others. When two different tribes met each other, conflicts were likely to arise and differences in social living, technological developments, or values came to light [10].

Today, in times of global convergence, these strong differences between tribes' realities and belief systems seem to disappear at first sight. However, due to social fragmentation, diversification, and the development of new communication channels in the field of information and communications technology (ICT), communities that establish themselves are not easily detectable. In the following, these communities are referred to as *virtual tribes*. Like ancient tribes, such virtual tribes define their own truths and live within their tribes' reality [9]. By using different tools, it is possible to identify and collect tribe members for any tribal macro-category which is the goal for an investigation by an analyst [12]. Later on, the likelihood of a certain social platform user being a member of one of these tribes can be measured by using machine learning techniques.

Holding more than 320 million active users [20] and 500 million *tweets* per day (Twitter, Inc.), *Twitter* is a great source of data that can be used for research. In the past, there have been lots of scientific investigations based on its plurality of accessible data, like extensive analyses for investigating the happiness paradox (friends in social networks generally seem to be happier than the considered user) or users' behavior on the online platform connected to income [1, 15].

While the access of information seems to rise in the progressing information era, people are able to hide behind their online accounts when indicating a statement of political or societal relevant nature. Investigating online accounts offers opportunities for data scientists to understand trends and sentiments of society and to draw conclusions on relevant character traits of online platform participants. In contrast to classical clipboard surveys, analyzing online accounts may mitigate honesty biases as people are more willing to disclose information in online environments [18]. Therefore, this approach allows a valuable complementary perspective on sensitive topics (political or societal) compared to results from a questionnaire. Findings can be used to guide decisions made by policy-makers in the real world as a person's personality characteristics and his/her behavior in both, real and online world, are significantly connected [6]. Findings depend on the respective chosen category of investigation. In our work, we chose to investigate controversies that arise around the topic of *sexuality*.

Sexuality encounters openness on the one hand and refusion on the other. Discussions about sexual orientation are shaped by the history and background of conflicting parties. Modern or traditional education and religious aspects influence the opinions of the panelists. Therefore, sexual orientation is a multilayered topic. Since the nineteenth century, organizations and communities have promoted a loosening of regulations against sexual orientations that are divergent to the conventional composition of a couple as man and wife [2]. Thus, they have made the discussion vivid and relevant to society. Disclosing communities that busy themselves with sexual orientation offers a better understanding of the composition of society as a whole.

Our work addresses the following research question: How do machine learning techniques allow us to conclude from users' behavior and language on Twitter to their attitudes about the *LGBT movement*? In order to answer this question, we first give the reader an overview about the theoretical background of our research and formulate four research hypotheses. Second, we explain our used methods in detail and reveal the results of our work. Finally, we critically discuss our findings and give an outlook for further research fields.

Theoretical Background and Related Research

This section will focus on discussing fundamental definitions the reader will encounter through the rest of this paper. Besides that, other related work will be briefly discussed in order to show the relevance of the topic.

Coins

Collaborative Innovation Networks (COINs) are innovation networks that are often self-organized and form independently of formal organizational structures in companies or within company networks [12].

Tribefinder

A tribe is "a network of heterogeneous persons linked by a shared passion or emotion" [4]. The system *Tribefinder* identifies these virtual tribes. Using data on the social media platform Twitter, it analyzes an individual's tweets by extracting information about key people, brands, used words, and topics of his or her tweets and categorizes the user into tribes belonging to five specific tribal macro-categories: personality, alternative realities, ideologies, lifestyle, and recreation. To analyze and identify the virtual tribes the continuous stream of tweets is an important source of information, which offers a powerful setting for studying and identifying tribes of individuals [19].

Using Tribefinder and the tribal vocabulary (which tribes are identified by which words or vocabulary) it learns, it is now possible to establish the tribal affiliations of every Twitter user. In practice, Tribefinder analyzes the individual's word usage in her or his tweets and then assigns the corresponding personality, alternative realities, ideologies, lifestyle, and recreation tribal affiliation based on the similarities with the specific tribal vocabularies.

Hypotheses

For the purpose of our research, we formulate four hypotheses. In order to clearly predict user's attitudes toward LGBT, we need at least two groups with different attitudes that differ in their language and behavior, which we call tribes of *LGBT* and *Anti-LGBT*:

H1: Two groups exist that highly differentiate in their attitude toward the LGBT movement.

H2: These two groups use different languages and reveal different honest signal characterizations.

In our work, we believe in the effectiveness of word analyses and demonstrate a bag-of-words approach:

H3: Analyzing users' words used in Twitter provides a high potential for prediction.

Finally, we apply our model to another tribe that consists of people who are against gun control regulations. Intuitively, we consider a convergence of opinions between the Anti-LGBT tribe and the contra-gun-control tribe as more likely than between the LGBT-tribe and the contra-gun-control tribe:

H4: There are more Anti-LGBT tribe classified people in the contra-gun-control tribe than the LGBT tribe classified people.

Methodology

To analyze large chunks of data, a proper framework or guideline is required in order to find the best amount of accurate data for our project. Since Data Mining is a creative process which requires different skills and knowledge, it is very hard to tie the success of the project to the knowledge of a single team member [16]. Therefore, we lean on the Cross Industry Standard Process for Data Mining (*CRISP-DM*) guideline which will merge our thoughts and guide us through a proper way of finding accurate data for the development of this project (Table 12.1). Many of the required steps and processes to gather the data have been discussed and addressed in section "Introduction" of this paper. The CRISP-DM model is divided in six phases which can interact in a cyclic pattern. The phases are categorized as follows: *Business Understanding, Data Interpretation, Data Preparation, Modeling, Evaluation,* and *Deployment* and will be discussed in this section [22].

For our project, we altered the first phase of the CRISP-DM model Business Understanding to *Domain Understanding*, since we are gathering and understanding

Table 12.1   Overview of process steps by software/methods used

Steps	Tools	Methods
Domain understanding	Twitter, Web-search	Reading, Discussion, Coolhunting
Data sourcing	Condor, Tribe Creator	Manual identification, Web search, Tribe creation
Data understanding	Condor, R-Studio	Statistical analysis methods
Data preparation	Condor, R-Studio, RapidMiner	Condor Preprocessing, R preprocessing
Modelling	RapidMiner	Model selection, process modeling
Evaluation	RapidMiner	Data Science evaluation measures
Deployment	Condor, R-Studio, RapidMiner	

information about a certain domain rather than a business venue. In our approach of Domain Understanding we worked around our main project task, which was to find out how different tribes with specific characteristics develop and correlate in digital networks. In order to do that, we brainstormed and gathered our ideas on which communities clash against each other the most and which ones were represented through a social media outlet such as Twitter. Out of this brainstorming session, we decided to analyze the correlations between LGBT and Anti-LGBT communities.

In order to discover more about the differences of the communities, we reached out to inform ourselves of the basic terminologies using *Google Scholar*, *Wikipedia* and implemented *Coolhunting* methods for identifying the most influential trendsetters of these characteristics. To find more information about Anti-LGBT and what it is comprised of, we looked for extremist groups and websites which promote this characteristic. We also started looking for representations of these communities on Twitter by identifying important and common *hashtags* and popular personalities within these communities. The gathered information out of *Data Interpretation* is discussed further in section "Results" of this paper.

We implemented a *Data Sourcing* phase before the Data Interpretation phase in our model which shows an alteration from the presented CRISP-DM model. At this point we used *Condor*, a software program developed by *Galaxyadvisors,* to measure the structure, content, sentiment, and influence of social communication networks over time. Condor also provides visualization features which we use to better understand the data we gather. Here we used three different approaches to collect the required data, which we derived as useful from the Domain Understanding phase. The first approach was focused on gathering the data via the Tribe Creator, also provided by Galaxyadvisors, where a certain keyword could be used as input such as a hashtag "#" in order to filter the results by the given input.

Here the tool would provide us with users and their Twitter ids, which we could use to search for friends and followers of that specific user. The second approach was to manually search Twitter, for specific users that would also use certain keywords, hashtags, or phrases. The third approach was to use Condor and its *tribe-fetch* function to find certain users who also used a certain keyword. With it we obtain a list of users which we then added to Tribe Creator. The focus of this phase was to create tribes (section "Introduction"), which we would later on use to create final datasets for our data mining model. The results of this phase will be thoroughly discussed in section "Results" of this paper.

In the Data Interpretation phase, we used the raw gathered data and implemented it in Condor in order to better understand the connection between every single actor. This phase will be closely tied to the *Data Preparation* phase, due to the functionalities and calculations that Condor provides. Thanks to the different visualization functionalities, the user can understand how different tribes differ in structure. Apart from that, Condor allows social network functionalities to be calculated such as degree centrality, betweenness centrality, and closeness centrality which all show the importance and position of certain actors within the network. The results of this phase will be presented in section "Results" of this paper.

In our Data Preparation steps, we used different tools to properly reduce the data for its optimal and most effective use. We decided that words, their frequency, and how often they appear within a certain tribe would help us to predict a certain tendency toward a tribe. Therefore, we needed to prepare the data in such a way that words should be the most resonant part of the data. In order to do so we first used some of Condor functionalities which calculate the *six honest signals of collaboration*, which are the most evident through the tweets we have collected through Twitter. The six indicators are namely *central leadership, rotating leadership, balanced contribution, rapid response, honest language,* and *shared context.* With these signals, future creativity, performance, and outcomes for teams can be predicted [13]. Besides the six honest signals of collaboration, another important way of making words the core of our data was to calculate the *Pennebaker Pronouns.*

Here the number of pronouns within a tweet of every user was counted. Condor has a built-in function that does so automatically and calculates the probability that a certain pronoun will appear in a tweet of the observed person [13]. Pennebaker discovered that how people use pronouns has a high predictive value [14]. After having Condor prepare the data, we exported it into an R Script which was written in the language R. This programming language is also an environment for statistical computing and graphics, due to its wide variety of statistics (linear, nonlinear classification, classical statistical tests, classifications and more statistical calculations) it seemed the most efficient solution for our data. With the R Script, we prepared the words in such a way that it can be used in a machine learning algorithm. The bag-of-words approach helps us in this specific task. The bag-of-words approach describes the occurrence or frequency of a word within a certain document [8]. Any other information besides the words are discarded. With the number of occurrences, it is intuitive that similar tribes will have similar words. This phase was tightly connected with Data Interpretation and the *Modeling* phase, since many iterations and changes to the data had to be made in order to fit it to our model.

In the Modeling phase, we decided to use an online modeling tool called *RapidMiner Studio*. RapidMiner Studio is a visual workflow designer, which helps develop prototypes for predictive models. Its graphical user interface (GUI) and documentation lead the user through the whole process of modeling and provide further information about every function, algorithm, or component that is used [17]. We integrated our prepared data into the tool and applied all predictive machine learning algorithms available in the toolset of RapidMiner Studio. Cross-validation provides solid accuracy metrics for a given model and its parameters. Based on accuracy comparisons of different model configurations, we selected the best choice of attributes, model, and parameters. After the Modeling phase, the *Evaluation* of the model is required. Here all results of the algorithms are taken under consideration. Our decision is mainly made by the highest accuracy provided by models which were calculated with different machine learning algorithms. Accuracy is calculated by the percentage of correct predictions over the total predictions. A correct prediction indicates that the value of prediction corresponds to the label attribute we specifically picked in the Modeling phase and is applied to the RapidMiner model. The results of our Modeling and Evaluation are discussed in section "Results".

It is important for us to develop a model which can be used for two scenarios. Firstly, for predicting a certain *tribe within another tribe*, and secondly, for predicting a *user's tendency* toward one tribe or another depending on his/her tweets. In the *Deployment* phase, we prepared the model in such a way that it is accessible for every example and dataset. This is achieved by a documentation of how to use the model and where to introduce the example dataset. To conclude our methods used during this project, it is important to understand the iterative and cyclic nature as seen in Fig. 12.1. Every phase can be altered in order to adjust the final dataset to provide the best possible outcome of the intended predictive model. Within the Deployment phase, the subphase *Demonstration* takes its place. A finished model used with real-time data tests its potential prediction.

Results

In this section, we will present the results structured by the phases of our adjusted CRISP-DM. We worked iteratively during the project making use of the loops the methodology provides. In order to provide a clear overview, we will only present the results of the last iteration of the respective phases here.

Domain Understanding

Sexualities are split up into several groups. There is heterosexuality which can be considered the most traditional and popular sexuality and describes the sexual preference for the, respectively, other gender. Besides, there are rather alternative sexual preferences such as homosexuality, bisexuality, transsexuality, and others. Finally, most alternative sexualities sum up in the LGBT movement. Therefore, we choose

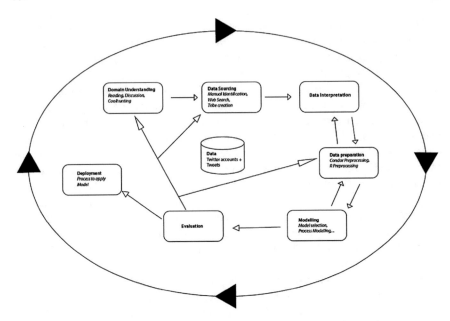

Fig. 12.1 Slightly adjusted CRISP-DM model with additional connections between processes that promote more flexible adjustments of particular process steps after the evaluation phase

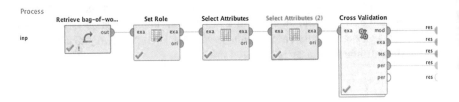

Fig. 12.2 Process in Rapidminer

accuracy: 77.47% +/- 6.78% (micro average: 77.43%)

	true V1 Anti–LGBT_483569...	true V1 LGBT_847ec453ee5...	class precision
pred. V1 Anti-LGBT_48356...	86	22	79.63%
pred. V1 LGBT_847ec453ee...	29	89	75.42%
class recall	74.78%	80.18%	

Fig. 12.3 Final model confusion matrix and accuracy

this group as a major tribe for our considerations. LGBT stands for lesbian, gay, bisexual, and transgender. Moreover, variants such as LGBTQ, LGBTQ+, LGBTQI+ exist, which is also reflected in hashtag usage. All these terms usually refer to the accuracy: 68.99% +/- 10.13% (micro average: 69.03%)

	true V1 Anti-LGBT_483569...	true V1 LGBT_847ec453ee5...	class precision
pred. V1 Anti-LGBT_48356...	82	37	68.91%
pred. V1 LGBT_847ec453ee...	33	74	69.16%
class recall	71.30%	66.67%	

Fig. 12.4 Model confusion matrix and accuracy without bag-of-words features

same community and the basic idea that open-mindedness toward sexuality is important and one should tolerate all sexual minorities. As a result of our Coolhunting we identified that LGBT is the most common hashtag and community that is referred to. Therefore, we defined our *LGBT tribe* as people who openly support lesbian, gay, bisexual, or trans. In order to get contrasting training data for our final model, we consider people who are significantly different from LGBT supporters. Therefore, we looked at people who are opposed to the LGBT movement. Typically, related keywords in the literature are homophobia and transphobia. In the course of our explorative research on Twitter, we identified a few potential subtribes regarding these attitudes. The spectrum reaches users on Twitter who are opposed to gay marriage to users who express in their tweets that alternative sexualities are diseases, that need to be cured, and users who verbally attack LGBT communities on Twitter in a disrespectful way. To include these different phenomena, we generally defined our *Anti-LGBT tribe* as people who are opposed to LGBT as sexual orientations.

Data Sourcing

Currently, the V1 LGBT tribe collected in Tribefinder contains 168 members who actively use Twitter. V1 Anti-LGBT consists of 119 members. The tribe-fetch with Condor resulted in two datasets of network (Twitter) data containing a total of more than 20,000 actors (users) and 480,000 links (tweets) including all the tribe members and their respective social networks on Twitter.

Data Interpretation

This stage was highly interrelated with the consecutive Data Preparation stage (see section "Methodology"). Therefore, we include results regarding features that were actually generated by the later Data Preparation stage. Apart from Condor generated features and visualizations, we look at the tribe member datasets resulted from Data Preparation including bag-of-words features. A look at the network graphs in Condor gives a first insight into the different tribes.

The graphs depicted in Fig. 12.5 display all actors and links to the respective tribes

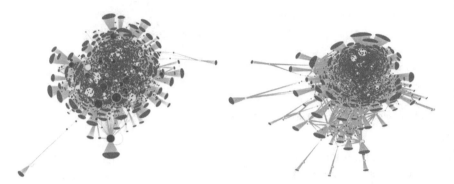

Fig. 12.5 Tribal network graphs of LGBT (left) and Anti-LGBT (right) tribes

in their surrounding network. The node color yellow highlights tribe members, the node size scales with the betweenness centrality measure. While both networks seem quite strongly connected, the LGBT network looks a bit dominant in this respect. Tribe members in LGBT are more often strongly connected and further in the middle of the graph. It is striking that the LGBT network is showing more non-tribe members that are quite central as well. In contrast, the Anti-LGBT network shows that a few tribe members are very central in the network (big yellow nodes) but there are a few central nodes of other tribes in the network. This could likely mean that the Anti-LGBT community is more isolated from and less connected with non-tribe-related important people. Moreover, there are mainly very central leaders and many non-central followers in the Anti-LGBT network.

The tendency of decentrality in the Anti-LGBT tribe versus collaborative centrality in the LGBT network is also reflected in the t-test results and boxplots (Fig. 12.6). The median Anti-LGBT tribe member has a lower betweenness and

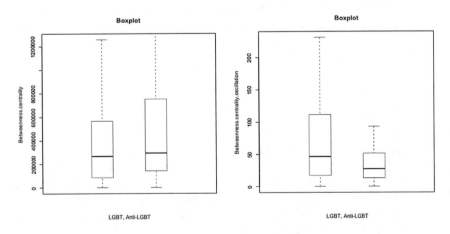

Fig. 12.6 Boxplots of betweenness centrality and betweenness centrality oscillation

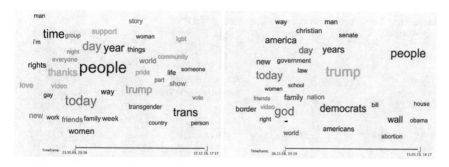

Fig. 12.7   Tribal word clouds of LGBT (left) and Anti LGBT (right)

degree centrality than the median LGBT tribe member. *Betweenness centrality oscillation* is relatively dominated by the Anti-LGBT tribe. The word clouds generated with Condor (Fig. 12.7) gives us a good feeling for the language use of our two different tribes. The size of terms depicts the relative frequency of terms in tweets. The color indicates the detected sentiment ranging from negative (red) to positive (green). It is obvious that the LGBT tribe has an overall more positive sentiment than the Anti-LGBT tribe (which is also confirmed by t-test results). Regarding the content, we find that Anti-LGBT tribe members significantly more frequently use political (e.g. "wall", "bill", "nation", "democrats", "senate", "Obama", and "Trump") and religious terms (e.g. "christian" and "god"). The language of LGBT tribe members is rather dominated by social terms (e.g. "community", "friends", "family", "people", "today") and LGBT related terms (e.g. "trans", "love", "pride", "gay", "lgbt", "transgender", and "person"). Regarding *Pennebaker Pronouns*, a look at the word usage distributions suggests that LBGT community members tend to use pronouns in a self-related way, if the pronoun is personal, while the Anti-LGBT community tends to use more non-personal pronouns—or personal pronouns linking to other people. In particular, the t-tests validate that "my", "me", and "it" are significantly more often used in the LGBT tribe. Anti-LGBT tribe members on the other hand significantly more often used the pronoun "the". Moreover, they use the pronouns "his", "they", and "that", coming as bag-of-words features, significantly more frequently (Table 12.2). These findings are also reflected in the weights of the final model's features, suggesting that pronouns features do well on contributing to the predictability of tribe membership (Tables 12.3 and 12.4).

Data Preparation

As a result of our Data Preparation, the final training and evaluation *dataset* is a combined dataset of all our tribe members with 134 features plus our target variable, the tribe name. 49 of the attributes come from the fetched actor data itself as well as metrics that are calculated from the actors' network by Condor (six honest signals of

Table 12.2 T-test results sorted by *p*-value, cut at $p \leq 0.05$

feature	p-value	lgbt_mean	anti_lgbt_mean
wall	2,73E+04	0.00181983330106428	0.0196835902088849
his	1,49E+05	0.0186030868611337	0.0385758966402175
presid	1,07E+06	0.0084632145513462	0.0266512267771265
democrat	1,48E+06	0.00488378672481424	0.023664798539182
will	8,01E+06	0.033411350465989	0.0581733192293614
they	2,96E+07	0.0285739054133454	0.0476930103709141
realdonaldtrump	4,40E+08	0.0179948844619235	0.0601055732114953
the	5,11E+09	0.418356562837061	0.522607922217094
peopl	0.000100533358739003	0.0525751221937414	0.0342322675309065
frequency_my	0.000135047097344263	0.00243479387222333	0.00126774235489974
frequency_me	0.000224745296962141	0.00199155660170316	0.00106036260005562
what	0.000458726842254082	0.0357170358214091	0.0495760177422029
should	0.000743342427618055	0.0140121480236884	0.020119864476807
Degree.centrality	0.00139158365662965	127.357.142.857.143	232.394.957.983.193
love	0.00172078485229054	0.035157306334108	0.0231379494996043
not	0.00179916978065713	0.0529034090616066	0.0682407490824305
statuses_count	0.00267150224494203	86.450.625	305.316.302.521.008
trump	0.00394782211251797	0.0198831201761341	0.0355391699705567
avg.sentiment	0.0055189834191805	0.517823002905533	0.479268834259697
whi	0.00726373382483415	0.016467524192345	0.021793280395551
look	0.00731389713599833	0.0233503410138929	0.0166119392092321
Betweenness.centrality.oscillation	0.00773359301712264	729.910.714.285.714	474.957.983.193.277
say	0.0106401217215972	0.02054332887485	0.025422998447025
get	0.0125406533127217	0.0337990085701445	0.0410704625509712
when	0.0131384930931877	0.0263026697629361	0.0334699716108629
know	0.01788637672863	0.0210117758058814	0.026167714922749
right	0.0189868852337561	0.0254718642426327	0.019000337684749
has	0.0215273577825814	0.0297930975778801	0.0368333844412554
avg.emotionality	0.0268918462896586	0.260084620104577	0.266640160515609
messages.sent	0.0289407993099403	47.872.972.972.973	727.822.033.898.305
this	0.0305944896060258	0.116576111161055	0.10169841927249
would	0.0345370635668361	0.0151652754556052	0.0199027446397517
frequency_it	0.0355062768340602	0.00322191758683079	0.00261660806975636
frequency_the	0.0396780822093489	0.0118753276060678	0.0134107995952259
Betweenness.centrality	0.0403804804570557	44.839.319.467.418	678.124.189.817.623
want	0.0431226000358793	0.0194807087847524	0.0237834510715019
Messages.sent	0.047790051320594	798.803.571.428.571	115.948.739.495.798
that	0.0500117914408883	0.0952595064308129	0.11075080516433

collaboration and Pennebaker Pronouns). Moreover, there are 85 attributes that are generated from the aggregated link data by means of our bag-of-words processing. It should be noted that the final training dataset consists of 115 Anti-LGBT entries and 111 LGBT entries due to filters in the process such as the filter in Condor that removes actors with much less activity for meaningful metrics.

Within the final modeling process in RapidMiner (see next paragraphs), we finally deselected some of the features: features with too many missing values as well as

Table 12.3 Features which the generalized linear model attributes to Anti-LGBT

Attribute	Coefficient	Std. coefficient t
Wall	−24.037	−0.516
Will	−12.100	−0.443
They	−13.207	−0.385
What	−10.200	−0.316
His	−11.678	−0.308
When	−11.418	−0.256
Presid	−8.830	−0.221
The	−0.883	−0.177
Democrat	−6.685	−0.173
Good	−9.214	−0.151
Intercept	0.676	−0.126
Whi	−7.068	−0.108
Avg.emotionality	−2.956	−0.068
Degree.centrality	−0.000	−0.053
Follow	−1.276	−0.049
You	−0.430	−0.047
Know	−1.715	−0.029
Betweenness.centrality	−0.000	−0.023
All	−0.568	−0.015
Not	−0.226	−0.009
But	−0.029	−0.001

identity-like attributes, such as names. This resulted in a final training dataset with 226 rows and 105 columns (features).

Modeling

Our model classifies a Twitter user as an LGBT (or Anti-LGBT) tribe member, given the entity including all its 104 features. Based on our Evaluation, we choose a *Generalized Linear Model*, a machine learning model for classification problems such as ours. We trained the model using the RapidMiner process (depicted in Fig. 12.2). The process consists of four data processing steps: data retrieval from the imported training data, selection of the target variable (Tribe), the final selection of features to be used for training, and the training and testing within a cross-validation (see Fig. 12.3).

Table 12.4 Features which
the generalized linear model
attributes to LGBT members

Attribute	Coefficient	Std. coefficient I
Peopl	13.720	0.490
Are	5.263	0.244
Right	11.311	0.236
Have	6.744	0.233
Look	11.171	0.212
This	2.583	0.136
Frequency_me	72.970	0.134
Betweenness. centrality. oscillation	0.002	0.126
Who	4.654	0.121
Frequency_my	52.673	0.119
Trump	2.552	0.106
New	2.732	0.081
Love	2.467	0.073
Avg.sentiment	0.592	0.063
Our	0.920	0.058
Frequencyjt	26.302	0.055
Vote	1.655	0.053
Frequency_have	49.927	0.051
Contribution.index	0.140	0.042
Frequency_was	17.980	0.023
Friends_count	0.000	0.007
There	0.349	0.006

Evaluation

The cross-validation of different feature sets and machine learning algorithms
revealed the best results for our final model, which utilizes a Generalized Linear
Model and 104 features. The final model's evaluation results are depicted in Fig. 12.3.
The performance can be summarized with 77.43% accuracy. The model performs
slightly more precise on Anti-LGBT predictions (precision: 79.63% versus 75.42%)
and slightly better recalls true LGBTs (80.18% recall vs. 74.78%). In other words,
if an actor is classified as Anti-LGBT it is more likely to be correct, and if an actor
is LGBT it is likely that he correctly gets detected as such, than it is respectively to
correctly classify an LGBT or detect every Anti-LGBT.

In order to decide on a specific algorithm, we tested six different machine learning
methods with RapidMiner Auto Model. It revealed that Naive Bayes and Generalized Linear Model performed best (Fig. 12.9). A follow-up analysis in the custom
RapidMiner process proved the Generalized Linear Model performs best for our final

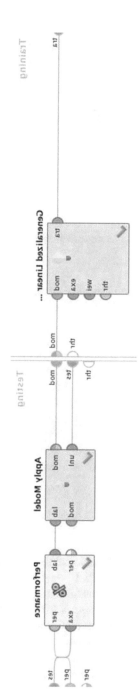

Fig. 12.8 Subprocess of cross-validation

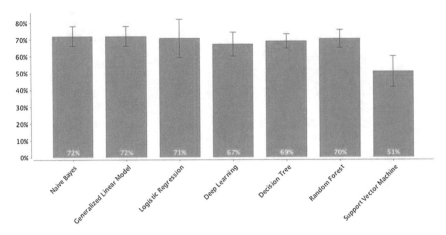

Fig. 12.9 Performance comparison of different classification methods with RapidMiner Auto Model

attribute selection cross-validation. Our evaluation also demonstrates the improvement caused by the inclusion of bag-of-words features. The cross-validation robustly shows that there is an improvement of around 8% (77.43% instead of 69.03%, Fig. 12.4).

The final configuration of bag-of-words specifies the maximal allowed sparsity parameter as 0.985. Words from messages are *stemmed* and *stop words* are not removed. Regarding this configuration, we did not evaluate all possible configurations, but took a look at different configurations within a reasonable range. Better results were reached with a higher maximal allowed sparsity level. However, we limited the allowed word sparsity at some point to keep the number of attributes relatively low. Stemming words and not removing words were proved to be dominant over all other combinations of these Booleans in terms of resulting model accuracy.

Deployment

One goal of ours evolved to be a deployable solution that allows the model application. To apply the model on new entities, we fetched the Twitter accounts of three single actors, namely Eminem, Donald Trump, and Peter Gloor. Moreover, we fetched another tribe, the *contra-gun-control* tribe. The application of our model yields the following results. Members of the contra-gun-control tribe are people who are supposed to like guns. According to our model they are mainly identified as Anti-LGBT (LGBT: 0.286 vs. Anti-LGBT 0.714). These single actors give us a good range of results. Donald Trump (*@realdonaldtrump*) is identified as an Anti-LGBT with a confidence of 97.6%. Marshall Mathers (*@Eminem*) is identified as an Anti-LGBT with a confidence of 73%. Finally, Peter Gloor (*@pgloor*) is identified as an LGBT with a confidence of 60.4%.

Discussion

Looking at our results allows us to draw conclusions to bolster our hypotheses. Results from our Domain Understanding indicate that there are at least two different groups, that highly differentiate in their attitudes toward the LGBT movement (H1). Our LGBT and Anti-LGBT tribes represent the two different groups that are attuned in either a positive or negative way toward the LGBT movement. Positions inside those groups can be (especially in the Anti-LGBT group) versatile in its level of aversion or affection. During the Data Sourcing we built two decent tribes by extensively manual inspecting every Twitter account for its veracity of attitude, that is desired for the respective tribe. Therefore, researchers can use the data from our tribes for further analyses as a solid fundament for their work. In the stage of Data Interpretation, we show that language and behavior differ between the members of the two tribes (H2). This manifests, for example, in sentiment and centrality measures and also in the word use of the tribe members. After proper Data Preparation, the results of our Evaluation phase indicate a high prediction potential for analyzing the used words by users (H3). Including bag-of-words features shows an improvement of around 8% in cross-validation. Interestingly, more general terms such as pronouns and conjunctions are shown to be more meaningful for our prediction value than more goal content-specific words. *Demonstration* inside our *Deployment* phase indicates that there are LGBT tribe classified people in the contra-gun-control tribe. However, the proportion of Anti-LGBT classified people in the contra-gun-control tribe is significantly bigger. Therefore, our hypothesis H4 can be obtained.

While we achieve strong results that are intended for satisfying support of our hypotheses, several limitations have to be taken into account. A rising quantity of data impedes the process of machine algorithm calculations and the preparation of sound prediction models, which leads us to limit the data quantity.

Nevertheless, concentrating on a limited quantity of data enables quality improvements like aiming at manually minimizing poor data as fake accounts and fake tweets, even if we do not emphasize nor quantify this procedure further. To improve the quality of our predictions we mainly focused on accuracy. We do not minimize the complexity of used features as we want to ensure a maximal accuracy irrespective of performance efficiency aspects. We do not investigate possible trade-off effects on accuracy and performance by limiting or adding different prediction features. In consideration of practicality aspects, we also consider the option of developing a more user-friendly IT-artifact as a proficient way for suitable applicability. Our present approach is more of a "do-it-yourself" one. Furthermore, our model is strongly attached to a certain domain. While it does well in the LGBT context, there is no proof that our procedure performs on the same level in other domains of use. Interested scientists could aim at diminishing the abovementioned limitations by elaborating on our research in further investigations. Also, we propose to expand the domain field of application to other areas. Applying our model to other tribes, like, for example, religion tribes, can provide insights into effects from tribe affiliations (like religious affiliation) on attitudes toward the LGBT movement. Our model also provides opportunities in the field of tailored marketing. Identifying a person's attitudes about a certain field can lay the foundation to create customized advertisements in the next step. Because this approach is likely to be manipulating, moral issues should be taken into consideration though. All in all, our work offers various insights into machine learning techniques for identifying attitudes from Twitter language and behavior plus a well-applicable model for the domain of the LGBT movement. While there is potential for further investigation, all of our previous formulated hypothesis can be obtained.

Conclusion, Outlook, and Limitations

In our work we showed how machine learning techniques allow us to conclude from users' behavior and language on Twitter to their attitudes about the LGBT movement. For this project we used an adjusted procedure of the CRISP-DM process. By identifying two groups of contrary attitudes toward LGBT, we created two virtual tribes by using the Tribefinder tool. We showed that language and behavior of users in the respective tribes differ. Furthermore, we identified word analyses as a valuable means of prediction. Thereby, specific terms are not as decisive as general ones like pronouns or conjunctions. Applying our model on the dataset of the contra-gun-control tribe reveals that the proportion of Anti-LGBT classified people in the contra-gun-control tribe is significantly bigger than LGBT classified people. The application of our prediction model on single Twitter accounts to identify a single users' attitudes toward the LGBT movement gives us comprehensible results. Further research could investigate how higher data quantities affect the model's quality. Furthermore, investigations could aim at applying our approach and model in different domains than the LGBT movement.

Appendix

See Figs. 12.6, 12.7, 12.8, 12.9 and Tables 12.1, 12.2, 12.3, 12.4.

References

1. J. Bollen, B. Gonçalves, I. van de Leemput, G. Ruan, The happiness paradox: your friends are happier than you. EPJ Data Sci. **6**, 497 (2017). https://doi.org/10.1140/epjds/s13688-017-0100-1

2. CNN Library, LGBT Rights Milestones Fast Facts (2015). https://edition.cnn.com/2015/06/19/us/lgbt-rights-milestones-fast-facts/index.html. Accessed 8 Mar 2019
3. Cambridge Dictionary "Tribe" (2019). https://dictionary.cambridge.org/de/worterbuch/englisch/tribe. Accessed 8 Mar 2019
4. B. Cova, V. Cova, Tribal marketing: the tribalisation of society and its impact on the conduct of marketing. Eur. J. Mark. **36**, 595–620 (2002)
5. D. Pritchard, Tribal participation and solidarity in fifth-century athens: a summary. Anc. Hist. 104–118 (2000)
6. D. Quercia, M. Kosinski, D. Stillwell, J. Crowcroft, Our twitter profiles, our selves: predicting personality with twitter, in *IEEE Third International Conference on Privacy, Security, Risk and Trust (PASSAT), 2011 and 2011 IEEE Third International Conference on Social Computing (SocialCom)* 9–11 Oct 2011, (Boston, Massachusetts, USA, 2011). Proceedings [including workshop papers. (IEEE, Piscataway, NJ, 2011), pp 180–185
7. Galaxyscope.galaxyadvisors.com.   https://galaxyscope.galaxyadvisors.com/tribe/donald trump. Accessed 11 Mar 2019
8. J. Brownlee, A Gentle Introduction to the Bag-of-Words Model. Abgerufen 14. März 2019, von (2017).https://machinelearningmastery.com/gentle-introduction-bag-words-model/
9. J.M. De Oliveira, P.A. Gloor, GalaxyScope: finding the "truth of tribes" on social media, in *Collaborative Innovation Networks* (Springer, Cham, 2018), pp. 153–164
10. M. Apelt, Forschungsthema: Militär: Militärische Organisationen im Spannungsfeld von Krieg, Gesellschaft und soldatischen Subjekten, 1. Aufl. VS Verl. für Sozialwiss, Wiesbaden (2010)
11. Merriam-Webster Dictionary "Tribalism". https://www.merriam-webster.com/dictionary/tribalism. Accessed 8 Mar 2019
12. P. Gloor, A.F. Colladon, J.M. de Oliveira, P. Rovelli, *Identifying Tribes on Twitter through Shared Context* (2018)
13. P.A. Gloor, Sociometrics and Human Relationships. Abgerufen von (2017). https://www.emeraldinsight.com/doi/abs/https://doi.org/10.1108/978-1-78714-112-420171027
14. J.W. Pennebaker, The secret life of pronouns. New Sci. **211**(2828), 42–45 (2011). https://doi.org/10.1016/S0262-4079(11)62167-2
15. D. Preoţiuc-Pietro, S. Volkova, V. Lampos, Y. Bachrach, N. Aletras, Studying user income through language, behaviour and affect in social media. PLoS ONE **10**, e0138717 (2015). https://doi.org/10.1371/journal.pone.0138717
16. R. Wirth, J. Hipp, CiteSeerX — CRISP-DM: towards a standard process model for data mining. Abgerufen 9. März 2019, von (2000). http://citeseerx.ist.psu.edu/viewdoc/sum-mary?doi=10.1.1.198.5133
17. Rapidminer, RapidMiner Studio-RapidMiner Documentation. Abgerufen 14. März 2019, von (2019). https://docs.rapidminer.com/latest/studio/
18. S. Benartzi, J. Lehrer, *The Smarter Screen: Surprising Ways to Influence and Improve Online Behavior* (Portfolio/Penguin, New York, New York, 2015)
19. S. Bringay, N. Béchet, F. Bouillot, P. Poncelet, M. Roche, M. Teisseire, Towards an online analysis of tweets processing, in *Database and Expert Systems Applications* (Springer, Heidelberg, Berlin, 2011)
20. Statista, Statista Number of monthly active Twitter users worldwide from 1st quarter 2010 to 4th quarter 2018 (in millions) (2016). https://www.statista.com/statistics/282087/number-of-monthly-active-twitter-users/. Accessed 8 Mar 2019
21. Twitter, Inc. Twitter für Unternehmen. https://business.twitter.com/de.html. Accessed 14 Mar 2019. Wikipedia.org https://en.wikipedia.org/wiki/LGBT accessed 06.03.2019
22. U. Shafique, H. Qaiser, A comparative study of data mining process models (KDD, CRISP-DM and SEMMA) **12**(1), 217–222 (2014)

Part III
Human-AI Interaction

Chapter 13
Digital Coworker: Human-AI Collaboration in Work Environment, on the Example of Virtual Assistants for Management Professions

Konrad Sowa and Aleksandra Przegalinska

Abstract Dominant opinion in the general public is that work automation will presumably hold negative societal implications, such as job loss, which often causes fear and misunderstanding. Contrarily to such an attitude, the approach we took in this paper is that people will experience rather positive effects of work automation, thanks to collaboration with artificial intelligence using virtual assistants. The quantitative experimental study was a business problem simulation. Participants were asked to perform tasks of a marketing manager in order to prepare a marketing campaign for a new product. Control group participants performed these tasks on their own, while experimental group participants did them in collaboration with a virtual chatbot-like assistant created specifically for this simulation. A total of 20 people participated in the study. A relevant difference in performance was observed between the groups, $n = 20$, $t(18) = 5.25$, $p < 0.001$. Participants collaborating with a virtual assistant achieved a 57% higher productivity (measured by tasks done) than those working on their own. Furthermore, in a post-study survery they assessed their productivity higher and were more satisfied with their performance. Results confirmed the hypothesis, proving that human-AI collaboration increased productivity within the studied sample.

All procedures performed in studies involving human participants were in accordance with the ethical standards of the institutional research committee with the Helsinki declaration and its later amendments or comparable with ethical standards.

K. Sowa (✉) · A. Przegalinska
Kozminski University, ul. Jagiellonska 57, 03-301 Warsaw, Poland
e-mail: konradsowa7@gmail.com

A. Przegalinska
e-mail: aprzegalinska@kozminski.edu.pl

© Springer Nature Switzerland AG 2020
A. Przegalinska et al. (eds.), *Digital Transformation of Collaboration*,
Springer Proceedings in Complexity,
https://doi.org/10.1007/978-3-030-48993-9_13

Introduction

Some fear the impact of widespread artificial intelligence (AI) on the labor market. However, much like the industrial revolution of the nineteenth century, the current *robotic* or *AI revolution* is believed not only to take away jobs from people, but also to create many new ones. Though the difference is that this time, office workers, rather than blue collars, are expected to experience the changes more severely. In a report from 2018 [16], BCG and MIT experts emphasized that some layoffs will be inevitable in the near future, and no large-scale job loss is expected, yet reskilling of the workforce will definitely be required. Ng [11] claimed that these are still the early stages of technology development. Calling AI "automation on steroids", he underlined a great deal of automation yet to come, however, with a neutral or even positive balance on job creation and job loss. This way of thinking also made its way to the general public, with Garry Kasparov, an international chess grandmaster, also known to be the "first man who lost to a computer" [8] saying that future is all about creating synergies between humans and machines powered by artificial intelligence. In this context, however, it is vital to ask what kind of impact on organizations and jobs should be expected. Is there any middle ground between jobs more suitable for humans and AI? Can people collaborate rather than compete with artificial intelligence? If so, how should such collaboration be designed for it to bring positive effects on productivity?

The goal of this paper is to explore synergies between human workers in managerial roles and AI-powered computer systems by experimentally verifying the assumed hypothesis. The underlying thesis behind the study is that a collaboration between humans and artificial intelligence in general (and virtual assistants in particular) increases productivity in management-related tasks.

Artificial intelligence is a research field focused on the design of computational agents with intelligent capabilities [14]. These capabilities include reasoning, knowledge representation, communication and language understanding, perception, learning, and others. There are various methods used to solve these problems, such as natural language processing, machine learning, or deep learning [9, 12]. The technology is widely used in business in applications like robotic process automation, chatbots, and virtual assistants or data mining.

With technological advancement, various types of interfaces came into existence. Artificial intelligence enables some of them, especially those based on communication the most natural for people—using natural language, voice, or gestures. Such interaction allows for a better simulation of more human-like interaction flow, leading sometimes to undesired outcomes, like repulse of fear, which in science is denoted as the Uncanny Valley Effect [4].

Further technical development and widespread application of artificial intelligence will doubtlessly impact how people live and work. Nonetheless, one must always ensure to keep people in the center of that development. Ethical and sustainable advancement is crucial when dealing with such a powerful technology. The goal should be to create environments in which both people and artificial intelligence can thrive together, utilizing what is best of each. Thereof, contrary to common opinion, this paper assumed collaboration rather than the displacement of people and artificial intelligence at work. More specifically, we focus here on collaboration on productivity.

In this study a business problem simulation was created. Participants were asked to perform professional tasks of a marketing manager and prepare a campaign for the launch of the new product. Participants of a control group did these tasks on their own and participants of the experimental group performed them in collaboration with a chat-based virtual assistant prepared for that purpose.

Literature Overview

This section summarizes key findings and trends related to the core topic: human-AI collaboration, particularly focusing on experimental research and papers focusing on the effects of such collaboration. The concept is fairly new, with most of the publications coming out in 2018 and 2019; thus it is safe to note that there is a scarcity of substantial research in this area.

An international group of researchers [5] developed a machine learning algorithm and signaling mechanisms, which were tasked with cooperating with humans in playing a selected set of five computer games. The results of a machine playing with a human were higher than in cases when people teamed up with other people. It has been proven that human-machine collaboration is achievable with the use of algorithmic mechanisms and furthermore, machines can be designed to learn and improve collaboration. An experimental study [18] exposed that productivity could be improved when the assembly system was specifically designed for collaboration of humans and robots. A robotic arm was constructed to support people in a simulated real-time assembly task using Lego bricks. A collaborative system achieved a 7% faster total assembly time and a 60% reduction in idle time in comparison with humans working by themselves. Moreover, Fugener et al. [7], created an experiment concerning image classification, in which the most productive results were achieved in a system where AI was delegating tasks to a human worker. Also, a recent report by MIT, presenting results of a global survey of business executives, has shown that a majority of respondents believed that AI could lead to a large increase of organizational productivity [16].

Methodology

The empirical study is an application of explanatory quantitative method based on an experiment. It aimed at verifying the assumed hypothesis about positive influence of human-AI collaboration on productivity. Such an outcome was expected from objective measures of productivity applied in the experiment, as well as from declarative measures collected after the experiment, provided by the experimental group. In a simulated environment, participants of experimental and control group were asked to perform tasks common for managerial roles. Participants of the experimental group performed those tasks in collaboration with a simplified chat-based virtual assistant created for that purpose, while participants of the control group performed the same tasks on their own. The third possibility, of a machine working on its own, was not tested, as—according to study assumptions—tasks had to include a human in the process. Productivity of each group was measured using time of task completion and quality of created outcomes.

Assumptions and Definitions

Artificial intelligence is a broad field of research. The technology has many different uses and to a various level has been adopted in organizations. Also, many tools currently applied in business already make a use of AI in the backend, as a supporting technology. It is, therefore, impractical to study the overall implications of human-AI collaboration. The focus of this study has been laid on the use of virtual assistants. Virtual assistants are a model example of how a natural human–computer interaction can be designed, as they provide an opportunity to communicate with machines using human language, directly in the user's environment, with the sole purpose of supporting humans in their errands.

According to the definition of the Cambridge Dictionary of English [2], synergy is "the combined power of a group of things when they are working together that is greater than the total power achieved by each working separately". That approach perfectly suited the assumed thesis, as well as provides a candid metric to be used in the quantitative study (comparing differences of effectiveness between the control group and experimental group). The synergy between people and AI can be understood as equivalent to a successful collaboration between them.

Productivity is a measure of efficiency and can be calculated by dividing generated outputs by invested inputs. A broader view considers productivity as an assessment of the efficiency of production. It entails the ability to convert resources into desired outcomes. In that sense, resources can be anything, like physical goods or time, alike outputs could be built products or achieved tasks. Individual productivity is a basic assessment of the employee's usefulness to the company. For a blue-collar worker, an example of a measure of productivity would be a number of products that the worker assembles within a workday, or an amount of resources he or she consumes to produce one product, or how much cripples does he or she generate [3]. The matter of white-collar workers productivity is much complex, as their work includes non-qualitative activities, such as management, teamwork, creative thinking, though for the simplified scenario in the study, it was assumed that, productivity of white-collar workers can be measured, for instance, by using time as input and quality combined with quantity as output.

Hypothesis, Variables, and Measurements

The key research question of the study was how does human-AI collaboration influence productivity in tasks related to management. Based on this question as well as on the presented literature overview and broad desk research, the following hypothesis has been formed: *collaboration between humans and artificial intelligence (virtual assistant) will increase productivity in tasks related to management among the studied subjects.*

In the experiment, the following variables were identified and measured:

- Dependent (response) variable Y—productivity (number of tasks done);
- Independent (explanatory) variable X—a collaboration of a subject with AI (group assignment).

A quasi-experimental method was applied to verify the assumed hypothesis. A level of the dependent variable (Y) was compared between two groups, where one variable (X) was altered in the experimental group and remained unchanged in the control group. The response variable was not measured in a pre-test. The control group was not affected by the explanatory variable ($X = 0$) and the experimental group was exposed to the variable ($X = 1$). Measurements of the response variable were taken during the experiment and compared between the two groups in subsequent data analysis. The following formula summarizes this method:

$$\begin{cases} E: X \rightarrow Y_E \\ C: \neg X \rightarrow Y_c \end{cases}$$

Explanation to Dependent (Response) Variable Y

The response variable of this experiment was productivity of participants measured by tasks which they performed in the simulation.

As explained in assumptions to the study, for white-collar jobs, both quantity and quality of output are relevant measures of productivity. Therefore, productivity can be measured using time as input and quantity combined with quality as output. Job completion can be measured, for example, by analyzing three aspects—time spent to do tasks, the number of tasks done, and how good the result was. The following formula can be applied to describe this reasoning:

$$productivity \overset{def}{=} \frac{output}{input} := \frac{quantity * quality}{timespent}$$

Participants of the experiment were asked to do tasks that are typical in managerial roles and their productivity was measured using time spent on those tasks, quantity of tasks done, and their quality.

While time and quantity can be represented by numerical values and remain completely objective, quality is a subjective measure to a high degree. An objective assessment of the quality of a performed task is possible, though this can only be found in tasks that are impartially provable to be done correctly or incorrectly. Such tasks include exam questions, determinations based on undeniable laws of science or nature, or those of which results can be seen directly after completion and numerically measured (e.g. some financial investment decisions). The aim of this experiment was, however, not to measure participants' knowledge of a particular functional area, but rather test them in a simulated management situation. In business, a given task cannot be easily and quickly assessed to be right or wrong, until enough time passes and the overall impact of its realization is estimated. A practical example here is a setup of a marketing campaign that cannot be directly assessed until a posterior market response (including unmeasurable outcomes like customers' affinity to a brand). Similarly, the accuracy of the HR department's decision on whom to hire could not be calculated. Because of this challenge, a certain assumption had to be taken regarding measuring the quality of outputs in the experiment. The quality of performed tasks was considered on the Boolean scale that is it could have taken a value equal to 0 or 1.

- Quality was considered to be 0 if not all of the requirements set forth in a given task were met.
- Quality was considered to be 1 if all of the requirements set forth in a given task were met.

If a given task was left unfinished or not all of the requirements of that task were met, the task was considered not to be completed. Specific tasks and their requirements were addressed further below. Participants of the experiment were made aware of tasks and their completion requirements. Participants were also informed before the experiment that their productivity will be measured.

For further reinforcement of the results of the experiment, a secondary response variable was adopted. It assumed measuring declarative productivity, which is the participants' own evaluation of the variable. In a brief post-experiment survey, the participants were asked to grade their productivity on a scale of 1–5. Averaged answers to this question were compared between groups.

Explanation to Independent (Explanatory) Variable X

A principal explanatory variable within this study was collaboration of people with artificial intelligence in the form of a virtual assistant in a work setting. It was denominated by a Boolean variable $\{0, 1\}$, categorizing participants as an experimental or control group. In the experimental group, participants used the virtual assistant to perform tasks, while members of the control group worked on the same tasks on their own.

There are, of course, other independent variables that could potentially affect productivity. As it was not a goal for this study to capture a plethora of productivity factors, these variables were considered either peripheral and their influence was measured and controlled, or noise variables that were eliminated. Selection of these variables was a result of the desk research [10, 13, 17]. Only variables that have a link to the simulation were accounted for. Variables that were connected with a particular organization were not considered (such as working culture, management style, recognition and awards, and team integration).

The following peripheral variables were measured:

- Tiredness (X_{p1})—this factor could not have been limited by the researcher, as it originates from various sources, such as stress, quality of sleep, time of the day, work being done before the study, and others. The level of this variable was measured in the post-experiment survey and its influence on the dependent variable was analyzed. Participants were asked to rate their tiredness on a 5-point scale.
- Focus (X_{p2})—this factor was partially limited by a setup of the experiment which forced participants to work in a separate and quiet room, with just a computer and tasks to do in front of them. Furthermore, all participants experienced comparable circumstances, therefore, the influence of external focus factors was alike across all of them. In the post-experiment survey, participants were asked to rate their focus on a 5-point scale, and the results of these indications were analyzed against the dependent variable.
- Task understanding (X_{p3})—the influence of this variable was limited by making the same onboarding process for all participants of the experiment and the same task management setup for all. Furthermore, in the post-experiment survey, participants were asked to indicate task understanding on a 3-point scale.

As far as noisy variables are concerned, the following ones were listed:

- Time pressure—in the experiment, the same type of time pressure was applied to all participants. They were informed about the time limit, however, there was no countdown visible. They only were informed when the time run out.
- Quality of tools—both groups were given the same toolset to perform tasks (apart from the virtual assistant in the experimental group). Overall productivity could have been impacted, however, the factor was leveled across all participants, thus no influence on the dependent variable was expected.
- Knowing the tools—proficiency in using working tools matters. For that reason, tools selected for the simulation had to be generally known and an assumption was taken that participants would be at least somewhat acknowledged with them. The influence of this factor was specifically expected in the experimental group, as people are not used to working with virtual assistants.
- Motivation—none of the participants were remunerated in any way for participation in the study; consequently, this could not have been a driver of their motivation. There also was no reward for completing a given number of tasks.

Recruitment Methodology of Experiment Participants

Participants were recruited to the study predominantly using the "interception" methodology [15, p. 32]. Candidates from varying communities and locations were approached and asked to participate in the experiment. Because the experiment was assumed to take approximately 40 min, in most cases sessions were scheduled. Due to technical limitations, and to limit a bias of participants influencing each other, only one person at a time could have experimented. Participants were not remunerated.

The following criteria were applied to recruit participants to the experiment:

- English skills—all materials provided as well as tasks and experiments conducted were in English, thus participants had to be fluent in the language. Participants' declarative responses from before the start of the experiment were taken as an assurance of fluency in English.
- White-collar job—during the experiment participants were asked to do tasks related to management. They did not have to be a manager at the time of the experiment, however, they should have had sufficient experience in an office job and had been exposed to managerial tasks.
- Marketing professionals or students were excluded—because the simulation was set in a field of marketing, people with a background in this area were excluded from the experiment. By virtue of their knowledge and experience, they would likely perform better than others and bias results of the experiment.

Experiment Flow

The experiment was a simulation of a business situation in which participants were asked to perform tasks requiring competences typical for managerial roles. These tasks had to embody skills applicable across different functional areas of business. Nonetheless, a functional area had to be selected. It ensured that the logical flow between tasks was kept and authenticity of the simulation elevated. A set of management competencies was generated through desk research and brainstorming, among others. It consisted of problem-solving, creativity, decision-making, social and communication skills, critical thinking, planning, leadership, delegation, goal setting, and time management. Maximizing usage of these skills was at the center point of the design of the simulation. Hence the following task selection criteria were applied:

- tasks put in use skills generally applicable in different managerial roles. This allows extrapolating results of the study to areas beyond marketing management, proving the usability of virtual assistants in all managerial roles, not only those in the one simulated in the experiment.
- there had to be a clear outcome coming from completion of the task, which would allow an objective verification if the task was completed or not. Tasks had to be possible to be done by humans and AI alone or in their collaboration, otherwise, the explanatory variable would not make any effect.

An analysis was conducted to examine different functional areas and potential problems to be solved by participants of the experiment. Drawing from that analysis, prototypes of different solutions were created and tested. A simulation of a marketing manager preparing a campaign for a new product was found to be most suitable.

Participants of the experiment were asked to put themselves in the role of a marketing manager at ABC Corporation. Their main goal was to perform preparatory works for the launch of a campaign for a new product that the company will sell. As a manager, it was their responsibility to set up an outline for the project and later lead it, upon approval from Chief Marketing Officer. The fictitious company—ABC Corp.— is a producer of coffee brewing equipment for households and offices. The new product to be launched on the market is an automatic coffee brewing machine. Tasks reflected steps typical to a marketing campaign preparation. They were set up in a logical flow so that one task could only be started if the previous was completed. There were nine tasks in the process, and 1 point was given for the successful completion of each. A task was only considered as completed when all required steps were fulfilled. It was not expected that all of the tasks would be done within the time limit of 20 min. This was acceptable because of points collected for each task, making up a productivity score of a given participant (Table 13.1).

The experiment started with a random assignment of participants to control or experimental groups. A coin toss method was used. It was followed with an introduction to the business simulation—the researcher briefly explained what is the simulation about, what are going to be participant's tasks, and what are the rules of their completion. Subsequently, tooling was presented and briefly explained, so that participants could focus their attention on performing tasks, rather than finding a way around tools they should use. The tools' setup was the same for both groups, except the virtual assistant, which was used only in the experimental group. Participants were asked to work on a Windows laptop and could have used any tools they found necessary to perform the tasks. The following simulated company system was prepared for them.

- Interactive Kanban board with task list and requirements (Trello);
- Spreadsheet with information about the company's finance (Excel);
- Simplified Gantt chart to be filled (Excel);
- Mailbox to send out emails and calendar blockers (Outlook Online);
- Cloud folder with all necessary files (OneDrive);
- Answer sheet to fill solutions to tasks (Google Forms).

After an introduction to tooling, participants were asked to go through a presentation introducing them to the simulation. In the presentation, the company was introduced, the participant's role was explained, there was an outline of existing products, and a business problem that the participant was supposed to solve was laid down. To keep a standardized task flow among all participants, the task list with fulfillment requirements was set up beforehand and presented. Time countdown started at the end of the presentation, on the approach to the first task. The experiment was finished either when participants finished all tasks or when they ran out of time. Participants were notified that there is a time limit of 20 min, however, there was no visible

Table 13.1 Tasks to be performed by experiment participants

Task 1—Market analysis	
Task description	Perform market analysis of competitive coffee machines
Required steps	1. Indicate market size of coffee machines segment (households + offices)
Competences	Analysis, critical thinking, prioritizing information

Task 2—Competition analysis	
Task description	Perform a competition analysis of coffee machines for home and office segment
Required steps	1. Indicate 3 key competitors of ABC Corp. in home and office coffee machines segment
Competences	Analysis, critical thinking, prioritizing information

Task 3—Advertising outline	
Task description	Create an outline to be used by copywriters in advertisements for the new product
Required steps	1. Analyze how competitors advertise their products in that segment; 2. List 3 things in which the new coffee machine is different from competition; 3. Write a short catchphrase promoting the new coffee machine
Competences	Creativity, abstract reasoning, analysis

Task 4—Team selection	
Task description	Assume ABC Corp. has a skill pool of people available for short-term projects. Their profiles are in the file management system. Select 3 people who will be part of your team for this project. You must have a graphic designer and someone to support you in marketing
Required steps	1. Select a graphic designer; 2. Select a marketing support
Competences	Social skills, planning, decision-making

Task 5—Project plan	
Task description	Create a list of tasks within the project and timeline necessary to complete them
Required steps	1. Think of 5 tasks that need to be done within the campaign; 2. Analyze how much time will be needed for each of these tasks; 3. Fill-in a Gantt chart from file management system
Competences	Planning, time management, organization

Task 6—Campaign budget	
Task description	Create a budget for the campaign
Required steps	1. Find out from the financial system what is the overall budget for new products marketing; 2. Calculate the total cost of the team; 3. Calculate how much budget you have left for advertising; 4. Decide on advertising budget split (TV/Press/Social media/Online ads)
Competences	Analysis, information search, decision-making, planning

(continued)

Table 13.1 (continued)

Task 7—Campaign goals	
Task description	Decide on the goals of the campaign
Required steps	1. Create 3 KPIs which will be used to measure success of the campaign
Competences	Goal setting, planning
Task 8—Campaign approval	
Task description	Summarize the campaign and ask your boss for approval
Required steps	1. Write an email with a summary of the campaign (all previous steps); 2. Send the email to Chief Marketing Officer of ABC Corp. to ask for approval of the campaign
Competences	Communication, social skills
Task 9—Kickoff meeting	
Task description	Schedule a meeting with the team to discuss the campaign and start working
Required steps	1. Create an agenda for the meeting (it should cover your findings from the preparatory process); 2. Find a date that suits everyone; 3. Send a calendar blocker
Competences	Information search, communication

countdown. That was to avoid the effect of them sacrificing the quality of the work just for fitting within the assigned deadline. An effect of the right balance between time pressure and work quality was sought. After the experiment, participants were asked to fill in a survey, whose purpose was to collect declarative data about using the virtual assistant (experimental group) and sociodemographic profile (both groups). The flow of the experiment has been presented in Fig. 13.1.

Virtual Assistant

In the experimental group, participants collaborated with a virtual assistant of ABC Corp., which was created specifically for this simulation. A simple chatbot was designed and developed using Chatfuel (https://chatfuel.com) and was accessible for participants via the Messenger app.

Task flow was known beforehand and the same tasks were assigned to all participants, thus most potential questions or commands were possible to predict and were the same for all users. This is a common practice in e-commerce or FAQ chatbots [6], where the majority of users ask similar questions, thus dialogue flow is alike among different interactions. The bot did not give direct answers to complex queries, nor did it perform whole tasks for people. Contrarily, conversations were arranged so that the assistant supports human worker, rather than doing work instead of him. He gave hints and solved tasks only partially and in cases where more information was needed from a human worker, he asked for clarification. To avoid any implicit bias to the participant–assistant interaction coming from the effects of the uncanny

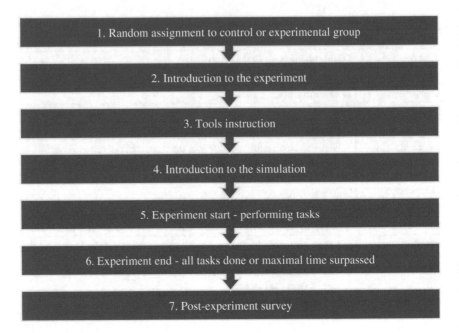

Fig. 13.1 Experiment flow

valley, the bot's design was dehumanized. That means, it did not have a name, did not express emotions, nor did it have any human-looking avatar. A short exemplary conversation with the ABC Corp. assistant has been presented in Fig. 13.2.

Post-experiment Survey

After the experiment, participants were asked to fill out a survey. Data gathered in the survey allowed to address certain limitations, as well as to measure an effect of peripheral or noisy variables on the experiment's results. Moreover, declarative indications of the influence of using the assistant on productivity were treated as a complementary data source for a result variable. Overall satisfaction from the performance of tasks was measured, as well as the satisfaction of collaborating the assistant. Sociodemographic profiles of participants were also gathered via the survey.

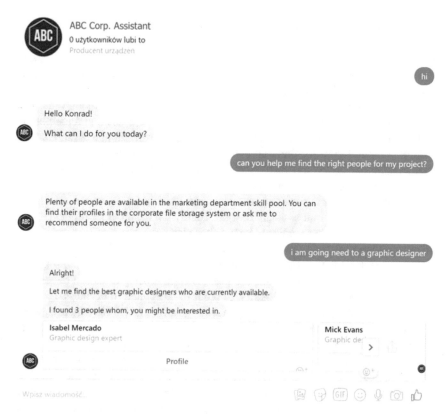

Fig. 13.2 Short conversation with ABC Corp. Assistant

Results of the Experiment

Limitations

While the experiment was conducted to the best of knowledge, skills, and availability of time and resources for the researcher, certain limitations still hold and are addressed in this section. Some originated from technical limitations of the software used to create the virtual assistant, other from a method of recruiting participants and the possibility of engaging their time. All these limitations should be addressed in future iterations of the experiment and its results verified upon various setups.

Chatfuel was chosen chiefly due to its usability, versatility, and speed of development of a bot. Additionally, its free version allows for up to 1,000 users of the bot, and for defining unlimited conversation scenarios. Despite the positive aspects, Chatfuel still holds certain limitations. For instance, Chatfuel allows only for text-based interaction, whereas an actual virtual assistant should allow for the multi-interface experience. Moreover, because of Chatfuel's simplistic NLP engine, the assistant was not able to handle a conversation outside of the simulated scenario. The software does not have sophisticated reasoning or learning capabilities, therefore all conversations had to be built from scratch. It is based on automated conversation flows that are launched by predefined keywords, rather than analyzing users' intent in a given context. It means that the bot is not able to answer a question formed in a way not similar to how it was defined in the backend, even though the question could be about the same thing. Not being able to understand and remember the context, it also does not support follow-up questions. Even though it did resemble traits of a natural conversation and some basic general queries were added (such as answering to "hello" or "thank you"), overall, it was limited to questions and commands referring to the simulated situation and tasks.

The purpose of this experiment was focused on outcomes of human-AI collaboration rather than different assistant designs. Surely bot design does influence interaction as proven by other studies [4], yet it has also been assumed that the positive effect of collaboration on productivity will hold regardless, and an optimal design could only make it better. For further studies on the topic, it is essential that more technologically advanced assistants are used. Different designs should be tested to optimize the interaction, and primarily, in order to be applied in business, the assistant should be versatile and capable of performing various tasks. Nonetheless, within the simulated environment, ABC Corp. Virtual Assistants performed or simulated performance of many functions that such assistants would typically have.

The experiment aimed at measuring people's productivity at work, however, it was conducted in close to laboratory conditions, where the work scenario was simulated, therefore intrinsically unnatural. Firstly, people are used to working with their own preferred room setup, desk, computer, operating system, mouse, and keyboard. Secondly, when they perform tasks in their job, they are well acquainted with the systems and tools that they use, whereas the experiment setup was new for them. The influence of this factor was assumed to be alike among experimental and control groups, therefore no differences between them were expected as its result, however, overall productivity could have been affected. This limitation was addressed in two ways—by using popular software tools for the setup, such as Windows 10, Microsoft Excel, Google Chrome, Trello, and onboarding process.

The onboarding process covered an introduction to the simulation and for that purpose, a presentation was used, explaining ABC Corp.'s business, the new product, participants' role, tasks, tools, and rules of the experiment. The second part was an instruction on how to use given tools, where to find information, and where to click to pass from one task to another. This process slightly differed between groups, as for the experimental group particular attention was given to explain how to use the assistant. Prior to performing tasks, participants were asked to introduce themselves to the assistant (by answering his questions—*How should I refer to you?*). They were also told to ask him for basic information about ABC Corp., like what does it do or about details of the new product they are supposed to advertise, and in this way, they had a short opportunity to practice asking questions and giving commands to the assistant. Furthermore, a help command was made available. It triggered the assistant to show examples of how to access his capabilities. It should be noted that for future iterations of the study, those participants should have more time to learn how to use the assistant. In the current version, the experiment took approximately 40 min per each participant, however, their longer engagement should be considered. In such a case, also the recruitment process and possibly the remuneration system should be altered.

Another limitation of the experiment is its scalability. To assure comparable results, all participants were provided the same working conditions (separate quiet room, a laptop with the additional monitor, and a simulated corporate system with a browser, files, and tools). Each experiment had to be done separately and took ca. 40 min to set up and execute. For further studies, a more scalable approach should be found, perhaps allowing multiple participants to work at the same time or for the online execution of the experiment. Otherwise, it would be resource consuming to undertake the experiment for a larger group of participants.

Participants

This subsection provides a set of descriptive statistics of the experiment's participants. The data was collected in a post-experiment survey and represents the sociodemo-graphic profiles of participants. A convenience sample methodology was applied to recruit participants.

There were 20 participants in the experiment. Out of that, 55% were male and 45% female. The age of participants ranged from 23 years to 44 years, with a mean of 28.9 years (SD = 5.43). There was an equal split between the experimental and control group, with 10 participants in each. The assignment to groups was random.

All participants had a degree of higher education, the majority with either a Master's degree (8 participants; 40%) or Master of Engineering degree (6 participants; 30%).

Participants were employed in a white-collar type of job. The majority of them did not hold a managerial position within their origination's hierarchy but were on a specialist (8 participants; 40%) or expert (7 participants; 35%) positions. Professional experience ranged from 1 year to 20 years, with a mean of 5.9 years (SD = 4.62).

Two market sectors of participant's current employment dominated. These were IT with 8 participants (40%) and telecom with 7 participants (35%). The question about the employment sector was open-ended and answers were grouped.

One of the key sociodemographic indicators was the participant's technology adoption. Answers were collected in the post-experiment survey. Before participants declared their adoption, a picture with a description of the technology adoption curve and its explanation [1] was shown. The achieved distribution reflects the curve, with the majority of participants being either early adopters (8 participants; 40%) or early majority (8 participants; 40%).

Another question asked in this part to all participants was about the frequency of using chatbots or virtual assistants. The median answer indicated a very rare usage of these technologies by participants (Me = 2). Most frequent answers indicated either never using a chatbot or virtual assistant (35%) or using it very rarely (40%).

Key Results

This subsection is aimed at presenting the results of the experiment, and the analysis of data is generated. A description of variables and their measurements were covered in section "Results of the Experiment", whereas here only the results were covered. Data was obtained from a measurement of the response variable of the experiment, which is a number of tasks done by participants, as well as from the post-experiment survey. For all participants, the same time limit of 20 min was set. It started as soon as they approached the first task, immediately after the introduction. Participants from the experimental group completed tasks in collaboration with a virtual assistant and participants in the control group performed the same tasks on their own. All participants reached a mean productivity score of 4.5 tasks, out of 9 tasks that were available. A significance level of 5% was assumed for all statistical tests (Table 13.2).

A mean result of the dependent variable (Y) in the control group was 3.5 tasks (SD = 0.71) and in the experimental group was 5.5 tasks (SD = 0.97). On average, participants of the experimental group did 2 tasks more than participants of the control group. The productivity of participants working with the virtual assistant was 57% higher than participants working on their own. The minimum productivity of participants in the control group was 3 tasks and the maximum was 5 tasks. For the experimental group, a minimum was 4 tasks and the maximum was 6 tasks (Tables 13.3, 13.4 and Fig. 13.3).

For verifying significance of the mean difference between groups, the following hypotheses were assumed and tested in a t-test for independent samples:

H_0: Productivity is equal in control and experimental groups.

H_1: Productivity is higher in experimental rather than in control group.

Levene's test confirmed equality of variances between groups (F = 1.2, p = 0.288). Results of independent samples' t-test proved statistical relevance, $t(18) = 5.25$, $p < 0.001$. The null hypothesis was rejected—participants collaborating with

Table 13.2 Summary of results for all participants

	N	Mean	Std. dev.	Min.	Max.	Median
Productivity (Y = tasks done)	20	4.5	1.28	2	7	4.5

Table 13.3 Summary of results for participants of control group

	N	Mean	Std. dev.	Min.	Max.	Median
Productivity (Y = tasks done)	10	3.5	0.71	3	5	3

Table 13.4 Summary of results for participants of experimental group

	N	Mean	Std. dev.	Min.	Max.	Median
Productivity (Y = tasks done)	10	5.5	0.97	4	6	5

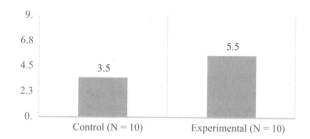

Fig. 13.3 Productivity results in control and experimental groups (number of tasks done)

artificial intelligence in the form of a virtual assistant achieved higher productivity. In a 95% confidence interval, the difference in productivity between using and not using AI was within the range of 1.2–2.8 tasks, corresponding to a 34–80% increase in productivity from an average score of a control group.

Supplementary Results

Further analysis was carried out for differences in the declarative rating of productivity between groups. It was assumed that productivity would increase not only in objective measures, but also in subjective indications of participants grading their results. In the post-experiment survey, subjects were asked to rate their productivity on a scale of 1–5. The mean productivity rating provided by participants of the control group was 2.6 (SD = 1.07), and by participants of the experimental group was 3.7 (SD = 0.67). On average, participants working with a virtual assistant rated their productivity 42% higher than participants working on their own. T-test proved relevance of this difference, $t(18) = 2.74$, $p = 0.013$. Levene's test confirmed equality of variances between groups ($F = 1.04$, $p = 0.32$) (Fig. 13.4).

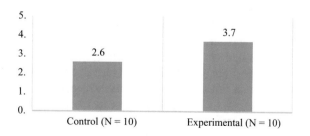

Fig. 13.4 Results of participants rating their own productivity after performing tasks in the experiment (number of tasks done)

Fig. 13.5 Results of participants rating satisfaction of their performance in the experiment (number of tasks done)

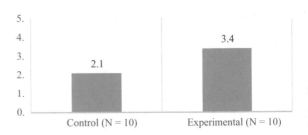

The third substantial difference between groups was the level of satisfaction of participants with their performance. Subjects were asked to rate how satisfied they were with their performance on a scale of 1–5. A mean satisfaction level in the control group was 2.1 (SD = 0.74), whereas in the experimental group it was 3.4 (SD = 0.97). On average, the satisfaction of performance was 62% higher for participants working with the virtual assistant. The difference was statistically relevant, $t(18) = 3.38; p = 0.003$. Levene's test confirmed equality of variances between groups ($F = 1.53, p = 0.232$) (Fig. 13.5).

Regression analysis of the results was not carried out due to several reasons. The decisive factor was the insufficient size of a sample. With 20 observations, a regression model would not bring satisfactory results. Furthermore, it was not a goal of this research to provide a function predicting productivity, but rather to analyze differences between studied groups and variables influencing those differences.

Influence of Peripheral Variables

In the subsection concerning the methodology of the experiment, three peripheral variables were mentioned—tiredness ($xp1$), focus ($xp2$), task understanding ($xp3$). To a large extent their influence was controlled, yet following literature review and some answers provided in interviews, they still could have impacted participants' results. For that reason, they were measured in the post-experiment survey and their influence against the dependent variable was analyzed.

Levels of tiredness and focus were considered to be denominated by values on an interval scale—they were participant's choice from a finite set of classifications, unequal to zero. A 5-point scale was used, with labels determining only minimum and maximum values (not at all; very much). Pearson's r correlation was selected as the most suitable. In the case of task understanding, the question indicated a 3-point Likert scale (Not at all; Somewhat; Yes), thus providing ordinal values as results. A statistical test—analysis of variance (ANOVA)—was undertaken to measure the influence of tasks understanding on the dependent variable.

Results of the analysis confirm that none of the peripheral variables had any significant influence on the result achieved by participants. Tiredness, $r(20) = 0.2, p = 0.397$, potentially could have had a stronger influence on productivity rather than

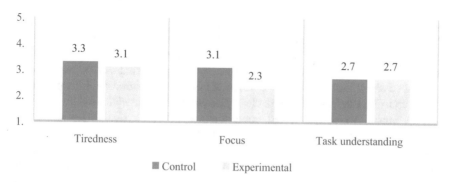

Fig. 13.6 Levels of tiredness, focus, and task understanding in control and experimental groups

focus, r(20) = 0.092, p = 0.700, yet both remained insignificant within the studied group. In-between group comparison of ANOVA for task understanding variable also showed insignificant influence of this variable on Y, F(1,18) = 0.27, p = 0.607.

Further analysis was carried out to verify if there was any difference in levels of peripheral variables between control and experimental groups. Analysis of descriptive statistics indicate a considerable difference in focus between groups—the focus was lower in the experimental group (M = 3.1, SD = 0.4) by about 35% than in control (M = 2.3, SD = 0.41), as it is visible on Chart 4. However, t-test for independent groups proved statistical insignificance of this difference, t(18) = 1.41, p = 0.176, as well as insignificance of difference on tiredness, t(18) = 0.37, p = 0.719, and task understanding, t(18) = 0.45, p = 0.660 (Fig. 13.6).

Even though it was not indicated in the literature, nor was it discovered in interviews, it was assumed that the technology adoption level of participants could have influenced their productivity, therefore, such a test was also undertaken. The technology adoption curve is represented by a normal distribution [1] and answers indicated by participants were ordinal, thereof an ANOVA was performed on the data. The influence of this factor over the dependent variable was insignificant, F(3,16) = 2.1, p = 0.140, however for further studies with larger samples, this variable should be taken into account, as its significance was fairly high along with a very high possible influence on the result.

Qualitative Results of the Experiment

The outcomes of the experiment were interesting, particularly in reference to the group of participants working with the virtual assistant. While the data does not provide a possibility of statistical inference, as it applies only for a small sample of 10 observations, certain conclusions can be formulated.

Firstly, an unquantifiable factor, which was the quality of the virtual assistant, could have influenced the productivity of participants of the experimental group.

Such dependency was suggested by participants of interviews and confirmed by participants of the experiment after the procedure was finished. However, in the post-study they indicated a rather positive experience—on a 5-point scale, a median satisfaction from collaboration with the assistant was rated at 4. Another question was asked about the participant's opinion of how the collaboration influenced their productivity and the answers were clearly positive, with a median of 4.5.

Participants were also asked about their opinion on five different matters concerning artificial intelligence. They answered by indicating to what extent they agree with a given sentence (Likert scale 1–5) and results were presented as medians of these answers. Again, most of the respondents indicated a positive influence of collaborating with AI on their productivity. Furthermore, they indicated that they would like to work with a virtual assistant in their job, even though their experience with the ABC Corp. assistant was only fairly positive. Most of the respondents said that they do not fear AI taking their job (Fig. 13.7).

The post-experiment survey also included three open-ended questions, which were intended to broaden a view of potential productivity factors, a discussion of which was started in the qualitative part of this study, as well as provide a set of potential subjective reasons behind participants' results. Results of positive and negative productivity factors were grouped and presented in Tables 13.5 and 13.6. The most common negative factor was that participants were new to marketing tasks, which, however, was one of the goals of the experiment—to expose all participants equally to a functional area which they did not have experience in. Other indications followed the factors anticipated earlier (Tables 13.5 and 13.6).

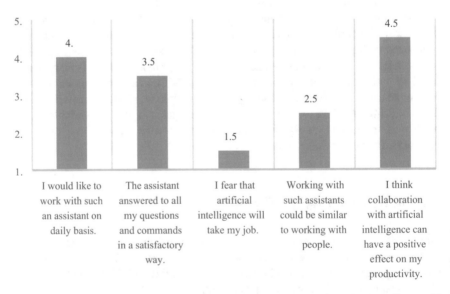

Fig. 13.7 Experimental group participants' opinion about working with artificial intelligence (N = 10)

Table 13.5 Negative productivity factors indicated by participants of the experimental group

Factor	Frequency of answers
New to these types of tasks	7
Knowing the tools	4
Time pressure	4
Experiment setup	3
Tiredness	3
Focus	1

Table 13.6 Positive productivity factors indicated by participants of the experimental group

Factor	Frequency of answers
Collaboration with the assistant	5
Experiment setup	3
Interest in the experiment	2
Time of the day	1

Another aim was to gather participants' feedback on working with the assistant. Below, a sample of answers to the question *How did you feel like when working with the assistant?* are provided:

- "Very handy tool for accessing data that otherwise needs to be manually searched on the web."
- "Chat is a good idea, collaboration felt natural."
- "I was more effective, calm and focused"
- "Okay, but it was something new to me, I never worked this way."
- "Like Tony Stark working with Jarvis."

Summary, Discussion, and Outlook

The hypothesis of increased productivity resulting from human-AI collaboration in the form of a virtual assistant was confirmed by the experiment on three levels. Firstly, a statistically relevant difference of dependent variable was observed between control and experimental groups. Secondly, a relevantly higher assessment of participants' own productivity was observed in the experimental group. Thirdly, qualitative results—a highly positive impression—was expressed by the experimental group participants about how collaborating with the assistant influenced their productivity.

Participants of the experiment working with a virtual assistant achieved a productivity 57% higher than their counterparts working on their own. The influence of other variables on the result was ruled out. It was confirmed by participants' own rating of their productivity, which was 42% higher for the experimental group. Not only users of the assistant achieved higher productivity, but another unexpected result was observed—overall performance satisfaction was higher by 62%. This indicates that collaborating with AI may allow people to have more time and focus on types of tasks which they prefer and in which they feel can deliver more value. Qualitative outcomes also bring compelling results. All experimental group participants left the experiment excited by working with the assistant. They liked the concept and would like to have such an assistant in their daily work.

While these results provide the desired outcome, limitations occurred due to the simplicity of simulation and technology used to create the virtual assistant.

Further studies should be undertaken to measure level of influence of human–AI collaboration on productivity in other scenarios. More sophisticated measures could be applied to both key variables—productivity and human-AI collaboration. In productivity, other disturbing variables should be accounted for statistically. Furthermore, a more thorough approach to task quality assessment is necessary. Non-binomial measures should be invented for collaboration, effectively allowing to build a regression model between variables.

Moreover, different forms of AI should be tested, with different assumptions and in different use cases. For this reason, authors urge a debate about formalizing a field of Human–AI Interaction, perhaps as a subfield of Human–Computer Interaction. It would allow for broad and focused studying of significant problems arising in this flourishing area.

References

1. J. Bohlen, G. Beal, The diffusion process. Special Report **18**(1), 56–77 (1957)
2. Cambridge Advanced Learner's Dictionary & Thesaurus, *Synergy* (Cambridge University Press, Cambridge, 2019)
3. B. Chew, No-nonsense guide to measuring productivity (Harvard Business Review, January 1988)
4. L. Ciechanowski, A. Przegalinska, M. Magnuski, P. Gloor, In the shades of the un-canny valley: an experimental study of human–chatbot interaction. Future Generat. Comput. Syst. **92**(03.2019), 539–548 (2018)
5. J. Crandall et al., Cooperating with machines. Nat. Commun. **9**, 233 (2018)
6. A.B. Ezra, Newstrail [Online]. https://www.newstrail.com/chatbots-now-taking-over-freque ntly-asked-questions-to-improve-engagement/ (2019)
7. A. Fugener, J. Grahl, A. Gupta, W. Ketter, Collaboration and delegation between humans and AI: an experimental investigation of the future of work (ERIM Report Series, Cologne, 2018)
8. G. Kasparov, Don't fear intelligent machines. Work with them (TED, Vancouver, 2017)
9. G. Kim, *Human–Computer Interaction Fundamentals and Practice*, 1st edn. red (CRC Press, Boca Raton, 2015)
10. A. Martin, The role of positive psychology in enhancing satisfaction, motivation, and productivity in the workplace. J. Organizat. Behav. Manage. **24**(1), 113–133 (2005)
11. A. Ng, AI for everyone. Coursera (2018)
12. P. Norvig, S. Russell, *Artificial Intelligence: A Modern Approach* (Pearson, New Jersey, 2010)
13. L. Peters, E. O'Connor, A. Pooyan, J. Quick, Research note: the relationship between time pressure and performance: a field test of Parkinson's Law. J. Occup. Behav. **5**(4), 293–299 (1984)
14. D. Poole, A. Mackworth, R. Goebel, *Computational Intelligence: A logical approach* (Oxford University Press, New York, 1998)

15. S. Portigal, *Interviewing Users—How to Uncover Compelling Insights* (Rosenfeld Media, New York, 2013)
16. S. Ransbotham et al., *Artificial Intelligence in Business Gets Real* (MIT Sloan Management Review, Massachusetts, 2018)
17. M. Rosekind et al., The cost of poor sleep: workplace productivity loss and associated costs. J. Occup. Environ. Med. **52**(1), 91–98 (2010)
18. L. Sayfeld, Y. Peretz, R. Someshwar, Y. Edan, Evaluation of human-robot collaboration models for fluent operations in industrial tasks (Rome, 2015)

Chapter 14
Collaborative Innovation Network in Robotics

Ahmad Khanlari

Abstract This study is the first attempt to employ Knowledge Building pedagogy and technology to integrate robotics into subjects like mathematics. Over the course of 6 weeks, 16 elementary students (Grade 5/6) engaged in engineering design process, computational thinking, and mathematical reasoning to design and program robots to collectively solve real-life issues such as creating a green and clean city. One of the knowledge building goals is to recreate schools as knowledge creation organizations. Therefore, this study employs the innovation network framework and uses social network and lexical analyses to analyze students' collaborations in robotics against knowledge creation organizations criteria and examine the extent to which student knowledge in math improved. The results show the emergence of the innovation networks in education settings and the importance of these networks for idea improvement.

Introduction

Since the mid-twentieth century, scholars have observed society gradually turning into a "knowledge society" (e.g., [10]). Prominent ideas running through both scholarly and popular commentary on knowledge society themes are *innovation* and *knowledge creation*. Therefore, helping students become more innovative and develop competencies associated with cultural creativity is an increasingly important goal in education settings [11]. This is in line with Knowledge Building theory and pedagogy, which aims to engage students in idea-driven knowledge creation and foster innovation and design thinking among today's learners [2]. Scardamalia and Bereiter [9] presented 12 principles that altogether describe knowledge building as a principle-based pedagogy that is set forth to make knowledge creation more accessible to teachers and students. These 12 principles frame knowledge building as an idea-centered pedagogy with students as epistemic agents, creating knowledge through

A. Khanlari (✉)
University of Toronto, 252 Bloor St. W, Toronto, ON, Canada
e-mail: a.khanlari@mail.utoronto.ca

© Springer Nature Switzerland AG 2020
A. Przegalinska et al. (eds.), *Digital Transformation of Collaboration*,
Springer Proceedings in Complexity,
https://doi.org/10.1007/978-3-030-48993-9_14

engaging in complex socio-cognitive interactions. Knowledge building places great emphasis on the notion of community and collective responsibility for knowledge advancement. In such a community, students are encouraged to advance community knowledge through participation in progressive discourse [1]. This is the community that Gloor [5] called Collaborative Knowledge Networks.

In an attempt to provide a framework to understand dynamics of knowledge creation communities and the contributors' behavior, Gloor [5] analyzed knowledge communities and identified three types of collaboration networks: Collaborative Innovation Networks (COIN), Collaborative Learning Networks (CLN), and Collaborative Interest Networks (CIN). These three networks together form the ecosystem of Collaborative Knowledge Networks (CKNs), which Gloor believes is an "extremely powerful engine of open and disruptive innovation for knowledge creation and dissemination" (p. 136).

Gloor described a COIN as a network of people that is formed by the most dedicated, self-organized, and intrinsically self-motivated people who have a collective vision. Although a COIN has the smallest number of members compared to the other two networks, it forms the core of a Collaborative Knowledge Network. Members of this network share ideas, information, and work in order to achieve the common goal. However, membership in COINs is fluid; founding members may leave the COINs due to changes in their interest.

Once a COIN is formed to create innovative ideas, other people join the community to discuss the new ideas emerging, to share a common interest and common knowledge, to learn about the purpose or the applications of the proposed ideas, and to collaboratively work to develop the ideas [4]. These people form a CLN; they not only—like experts—actively share knowledge but also—like students—actively seek knowledge [5]. The members of this network include people who may lack the skill, time, or interest to join the COIN, but are interested in discussing the ideas emerging and the knowledge being shared [5]. The CIN network is formed by people who typically "do little actual work together in a virtual team" [5]. While a minority of people in this network share knowledge, the majority of them are silent knowledge seekers (lurkers) who do not usually contribute any content [5]. Each of these three networks is embedded into the subsequent, larger community. The COIN is the smallest community which is the heart of this set of concentric communities, while the CIN is the largest and outermost community.

This study explores how and if employing Knowledge Building pedagogy and technology cultivates and facilitates knowledge creation in an elementary class where the students explore math using robotics. While educational robotics often goes on in a competitive way, this research employs Computer-Supported Collaborative Learning (CSCL) environments to create an open innovation community. Collaborative innovation network framework [5] is employed to analyze a knowledge building community according to the innovative networks' dynamics, to explore whether students have similar contribution patterns as those observed in innovation-intensive organizations.

Method

This study explores forms of engagement and knowledge work within a knowledge building class. Math topics (e.g., measurement, proportion, and conversion) are explored using robotics. The online environment used to facilitate the formation of innovation network is Knowledge Forum® (KF)—a web-based discourse medium specifically designed to support production and refinement of community knowledge to advance understanding of the world and effective action through social interaction.

Participants and Settings

Participants of this study included 16 Grade 5/6 (12 boys, 4 girls) in a school located in Ancaster, Hamilton, Canada. This school has two classes per grade, taking into account that in some classes two grades are mixed. In this study, the participants explored math concepts while working on their robotics projects over the course of 4 months, one session a week. Each session lasted for 90 min, involving two components:

– Knowledge building talk in which students were engaged in face-to-face discourse in order to update their peers about their progress on the task, express their success/failure stories, ask questions, and answer their peers' questions. The knowledge building talk usually lasted for 20 min. Students were asked to enter a summary of their knowledge building talk into Knowledge Forum.
– After each knowledge building talk, students were divided into their groups to work on their projects. During their project time, students were involved in hands-on robotics activities to solve the challenge which lasted for an hour. While working on their projects, students were expected to enter their findings, challenges, issues, and breakthroughs to Knowledge Forum, ask their questions, and build on other students' contributions. During all the two components described above, minimal guidance to students was provided.

Dataset

Students' contributions to Knowledge Forum and the log data are the two primary sources of data for this study. Students were involved in three different projects; they posted 199 notes for project 1, 107 notes for project 2, and 163 notes for project 3. While students' interactions during all the three projects have been analyzed, this study only presents the results of the final project because the methods of analysis and the procedures in all the three projects are similar. For the final project, students decided to design a robot to collect the garbage from the environment and make their city a "green city." As part of this project, students were expected to consider math concepts when designing their robots. During the study, the teacher and the researcher purposely provided no guidance and let the students think and decide about the projects, in order to turn agency over to students.

Plan of Analysis

The data analyses include social network analysis and lexical analysis [3, 8]. In order to be able to explore changes in students' collaboration patterns, students' notes posted during the project are divided into two parts: the first half and the second half of the project, each part lasted for 3 weeks. In each half part, a social network analysis was first conducted in order to examine if the three subnetworks of the knowledge networks (i.e., COIN, CLN, and CIN) are formed in a community, and to identify the members of each network. Then, a lexical analysis is employed to explore whether new ideas were proposed and spread in the community. To further analyze students' contributions, we calculated the "math-lexical density." Lexical density is defined as the proportion of content words (nouns, verbs, adjectives, and adverbs) to the total number of words. In this study, I am especially interested in math words, therefore, to calculate the "math-lexical density," I only selected math-related words and calculated the math-lexical density for each part of the modules. To calculate the math-lexical density, the total number of math terms used by the students are divided by the total number of words. This measure can show what percentage of the words students used in their notes are math-related words. To further analyze students' notes, all the notes posted by students are reviewed and categorized into two categories: math notes and non-math notes. A math note is a note that contains at least one math-related keyword. Then, the percentages of the math notes in each part of the modules are calculated and labeled as the "math-note density." These two analyses can reveal the breadth of student math talk. I conducted such analyses over time to examine how students' discussions around those math concepts change over time.

Results

This section details the results of the analyses, including a description of findings for the social network analysis and lexical analysis.

Social Network Analysis

One primary objective of this research was to explore whether the three networks observed in knowledge creation organizations are formed in education settings. To

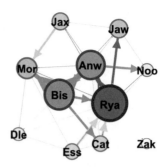

determine if the three subnetworks are formed, social network analysis was employed
and students' collaboration patterns have been analyzed over time.

Part 1

Figure 14.1 shows the networks of the writers in the first half of the module.

According to Gloor [5], a central dense cluster of the network is an indicator of
the emergence of a COIN. Gephi, the social network analysis tool that is used in
this study, has a feature to distinguish dense nodes using different colors and sizes.
A dense node is a node with a high number of lines (in-degree and out-degree)
incident to it. In Fig. 14.1, nodes with higher degrees are presented in dark green,
with larger sizes. As the figure shows, Rya, Anw, and Bis form a central dense cluster
in the network, which may be an indicator of the formation of a COIN. To verify the
formation of the COIN, the whole network is divided into two parts: the potential
COIN and the potential CLN/CIN (Fig. 14.2).

Gloor described density and Group Betweenness Centrality (GBC) measures to
distinguish COIN members from CLN/CIN members. Network density is defined
as the ratio of the *actual* number of communication connections to the maximum
potential communication connections [3], and reflects the extent to which students
interact with each other. The density of each subnetwork is calculated using Gephi.
Also, the group betweenness centrality of these two networks are calculated using
Freeman's index:

(a) |**(b)**

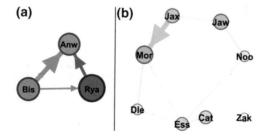

Table 14.1 Density and GBC for the networks in Part 1

Network	Density	Group betweenness centrality
COIN	0.83	0.06
CLN/CIN	0.18	0.28

$$BC_{group} = \frac{\sum_{i=1}^{g}[BC_* - BC_i]}{(N-1)} \tag{1}$$

Freeman (1979) has defined the group betweenness centrality index as a measure of the homogeneity of the members' betweenness; "lower betweenness centrality means the communication behaviour of the group members is more egalitarian" [6]. The results of the analyses are presented in Table 14.1. As Table 14.1 shows, the identified COIN has the highest density and lowest group betweenness centrality, while the identified CLN/CIN has the lowest density and the highest group betweenness centrality. These results confirm that the COIN and CLN/CIN networks have been correctly identified. To identify the members of each team, the log data are used.

Figure 14.3 represents the log data for note creation. As described by Gloor, the CLN members are active contributors who actively engaged in sharing knowledge, while the CIN members are usually silent knowledge seekers who make little or no contributions. Therefore, those who have posted notes more than the average number of notes posted by the CLN/CIN members form the CLN network, while other students who posted fewer notes than the average form the CIN network.

As Fig. 14.3 shows, Dle, Ess, Jaw, and Jax have written more than the average number of notes posted by all the CLN and CIN members. Therefore, they are

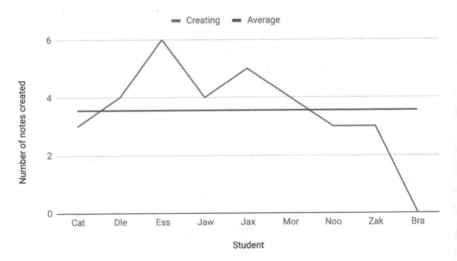

Fig. 14.3 Number of notes created by the CLN/CIN members in Part 1

considered as CLN members. As a result, Cat, Mor, Noo, Zak, and Bra are considered the members of the CIN.

Part 2

Figure 14.4 shows the network of the writers in the second part of the final project.

As it is evident from Fig. 14.4, Rya, Bis, Anw, Zak, and Dle can be considered as the potential COIN members, because they together formed a central dense cluster. To verify this selection, the network is divided into two subnetworks: the potential COIN network and the remaining part that potentially form the CLN/CIN network (Fig. 14.5).

As Table 14.2 shows, the COIN and CLN/CIN networks are correctly identified because the identified COIN network has the highest density and the lowest Group Betweenness Centrality.

Similar to the previous part, to identify which students are members of the CLN and which students form the CIN, the log data are used (Fig. 14.6).

As can be seen in Fig. 14.6, Cat, Ang, Ess, Jaw, and Jax have posted fair numbers of notes (i.e., above the average number of notes posted by all the CLN/CIN members).

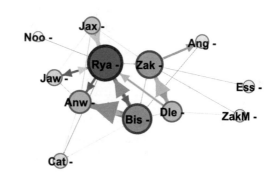

Fig. 14.4 Network of writers in Part 2

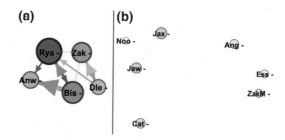

Fig. 14.5 **a** Potential COIN network, **b** CLN/CIN networks in Part 2

Network	Density	Group betweenness centrality
COIN	0.7	0.08
CLN/CIN	0.048	0.21

Table 14.2 Density and GBC for the networks in Part 2

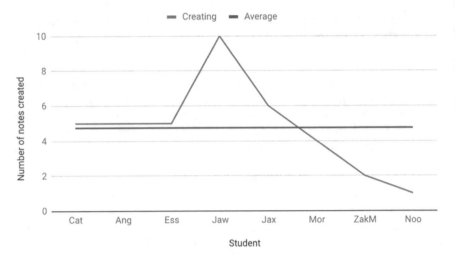

Fig. 14.6 Number of notes created by the CLN/CIN members in Part 2

Therefore, they form a CLN network, and the remaining students (i.e., Mor, ZakM, and Noo) are considered as the members of the CIN network.

Lexical Analysis

Using the lexical analysis tool embedded in Knowledge Forum, I examined the math words students used in each part of the project. In this section, the results of the analyses for both Part 1 and Part 2 are presented.

Part 1

Figure 14.7 shows the math words students used and their frequencies. From Fig. 14.7, it is evident that students mainly discussed *cardinal directions* as they have written the terms cardinal directions (7 times), east (6 times), West (5 times), north (4 times), and south (3 times). They also talked about other concepts like measurement and graphs. However, as the frequency of the used terms shows, these concepts have not been well-discussed since these concepts were discussed only a few times.

The math-lexical density and math-note density of the first part are calculated as 5.75% and 30%, respectively. Therefore, less than 6% of all the words students used in their notes are math-related terms. On the other hand, the math-note density shows that only 30% of the notes (23 notes out of the total 76) contain at least a math word. The results show that while students discussed some math concepts, they were mainly discussing non-math-related issues, such as building their robots. For example, one of the students had written:

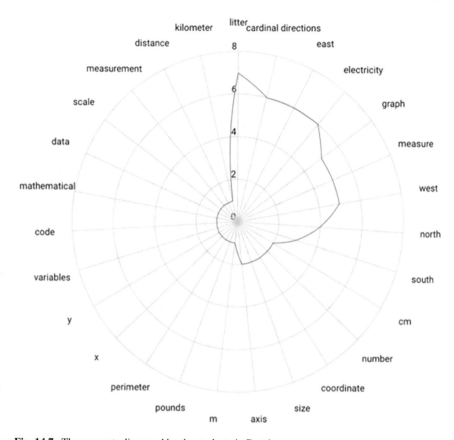

Fig. 14.7 The concepts discussed by the students in Part 1

We had a problem when the vex [their robot] turns and stops. We tried cutting the bottom and that didn't work. We tried to fix the wires and that didn't work.

As the next step, I reviewed the math notes to examine who wrote those math notes and who initially introduced those math words. Interestingly, most of the math notes are written by the COIN members; out of the total 23 math notes, 16 were posted by the COIN members. Also, most of the math terms were first introduced by the COIN members. For example, "cardinal directions," "coordinate," "axis," and "scale" were first introduced by the COIN members. For example, Anw was the first person who discussed cardinal directions:

We would try to use cardinal directions (north, east, south, west) and then try to make a way that the vex could use to go around using seconds.

However, there were other math concepts that were initially introduced by a student from CLN/CIN networks. For example, Dle, who is identified as a member of the CLN network, was the first person who used the word "pounds":

My theory is… we can measure the total garbage at the end of the week. We can use pounds to measure.

Part 2

Employing a similar approach, a lexical analysis has been conducted on students' notes posted during the second part. Figure 14.8 shows the math words students used and their frequencies. As can be seen, students discussed different math concepts, such as "measurement," "volume," "area," "distance," and "cardinal directions." The frequencies of the topic-specific words that students used show that students deeply discussed those topics. For example, students extensively referred to words related to volume (e.g., ml: 17 times; volume: 7 times; oz: 3 times). They also well discussed concepts related to cardinal directions, distances, and areas.

In the next step, the math-lexical density and the math-note density are calculated. In this part, students used 364 math words out of the total 1690 words (21.53%).

Fig. 14.8 The concepts discussed by the students in the Part 2

Also, out of the total 84 notes posted by students, 65 notes (62%) were coded as math notes. The following note is an example of the math notes:

> We thought of doing one space in the hallway, this is how we got the area. First, we measured the length and width of a square tile. That was: 30 and 30. Then, we multiplied 30 by 30 that equals 900. Now we counted how many square tiles there were, that was 80. Finally, we multiplied 80 with 900 and got 72,000.

On the other hand, in the non-math note, students mainly discussed the codes they wrote for their robots. To further analyze who wrote the math notes and whether the math concepts were first introduced by the COIN members, I reviewed all the math notes. The results of the analysis show that, out of the total 65 math notes, 43 nodes were written by the five COIN members. Also, most of the math terms were first introduced by COIN members. For example, Zak, who is identified as a COIN member, was the first person who talked about the area and how the area of a rectangle can be calculated:

> ... we measured the side lengths. The width we got was 160 cm and the length we got was 192 cm. So our space is obviously a rectangle. After that, we wanted to find the area so we did the area formula Length times width and so now we multiply $1.92 \text{ m} \times 1.60 \text{ m} = 3.072 \text{ m}^2$.

However, there were some math concepts that were initially introduced and discussed by other members. For example, Ang, who is identified as a member of the CLN network, was the first student who talked about the perimeter:

> Are you sure you did multiplying because the way you showed it looks like you added which is the formula for the perimeter?

Discussion and Conclusion

As the results of the analyses show, the three networks observed in knowledge creation organizations are formed in this Grade 5/6 class. Interestingly, not only the COIN, CLN, and CIN are formed in both parts, but also students moved from one network to the other two networks. For example, while Bis, Rya, and Anw formed the COIN in the first part, the COIN members in the second part included Bis, Rya, Anw, Zak, and Dle. As discussed, Dle was initially a member of the CLN network, and Zak was a member of the CIN network. Also, Cat from the CIN moved to the CLN network. As it is evident from the results, compared to the first part, more students were involved in the community: in the second half of the project, more students formed COIN, CLN, and CIN. As the results show, while in the first part only three students formed the COIN, in the second part five students formed the COIN. This result is supported by the literature as Gloor [5] stated the COIN network might extend over time to include other members. Also, in the second part, five students formed the CLN network, in contrast to the three students who formed the CLN network in the first part. As a result, while the CIN in the first part composed of four students, in the second part its members reduced to three. The results confirm the emergence of "leaders" and "rotating leaderships" [5] among students; as stated by Gloor [5], during the creative "swarming" process, several leaders emerge as the community is a self-organized community aiming to advance their goals. In essence, highly creative teams have different leaders who frequently rotate over time [6, 7]. The results of this study show that, over time, different students took the role of "thought leaders" to move the discussion toward knowledge objectives. These thought leaders take collective responsibility for contributing to the community knowledge, resulting in the success of the project.

Moreover, the results show that while in the first half of the module new ideas and new concepts were introduced (mainly by the COIN members), they did not spread. For example, the words "Coordinate," "Axis," "Perimeter," and "Area" were only used by the COIN members. Some other words like "Pounds" were discussed only by a few students. This result was expected as most of the students were not actively engaged in the community. As the log data show, several students in this part did not even read notes. The average number of notes read by CLN/CIN students was 44.5 notes per person, which may not be enough to learn the new concepts introduced in the community. Also, the average number of notes posted by CLN/CIN members was 3.5 notes per person, which may not be enough to spread the ideas. Also, only three students formed the COIN, and this may have affected the spread of their ideas.

In the second part, compared to the first part, not only more diverse math concepts have been discussed, but also some concepts like measurement, area, and cardinal directions have been deeply discussed in the community. Such a result was expected because the total math notes increased from 30% in the first part to 62% in the second part of the project.

Interestingly, the results show that the majority of the math notes in both parts are written by COIN members: in the first part, 65% of the math notes and in the second part 66% of the math notes were written by the COIN members. Such findings indicate that the improvements that happened from Part 1 to Part 2 were a result of the great effort that the COIN members put to move the discussion forward. In essence, not only most of the new math concepts were first introduced by the COIN members, but also they were mainly discussed by the COIN members. This result shows the importance of the COIN members for knowledge advancement in a community. However, the contributions of other community members (i.e., CLN and CIN members) should be acknowledged, as all these three networks together form the ecosystem of knowledge creation.

This research provides an empirical study where we employ Knowledge Building pedagogy and technology in robotics. As this research shows, it is still possible to benefit from robotics while students engage in whole-class collaboration. Therefore, Knowledge Building pedagogy and technology provide an alternative approach to competition-based learning which is a common approach in educational robotics. Employing Knowledge Building pedagogy and technology not only lets students take advantage of the great opportunities that robotics provides, but also can improve their collaboration and communication skills, which are considered as twenty-first century competencies. Such analyses can be conducted in education settings in order to examine the extent to which students' collaboration changes over time as a result of being engaged in community knowledge. One of the limitations of this study is that no quality analysis has been conducted to qualitatively analyze students' contributions; rather, only lexical analysis has been conducted to examine the community in terms of knowledge creation. While the lexical analysis provides insights about the knowledge building discourse, it is unable to reveal whether the community is moving toward the knowledge objectives, which is idea improvement. This limitation will be addressed in the next study.

Acknowledgments All procedures performed in studies involving human participants were in accordance with the ethical standards of the institutional research committee with the 1964 Helsinki declaration and its later amendments or comparable with ethical standards.

References

1. C. Bereiter, Implications of postmodernism for science, or, science as progressive discourse. Educ. Psychol. **29**(1), 3–12 (1994)
2. C. Bereiter, M. Scardamalia, Education for the knowledge age, in *Handbook of Educational Psychology*, 2nd edn., ed. by P.A. Alexander, P.H. Winne (Lawrence Erlbaum, Mahwah, NJ, 2006), pp. 695–713
3. A. Degenne, M. Forsé, *Introducing Social Networks* (SAGE Publications, London, 1999)
4. M. De Maggio, P. Gloor, G. Passiante, Collaborative innovation networks, virtual communities, and geographical clustering. Int. J. Innov. Reg. Dev. **1**(4), 387–404 (2009)
5. P. Gloor, *Swarm Creativity: Competitive Advantage Through Collaborative Innovation Networks* (Oxford University Press, NY, 2006)
6. P. Gloor, R. Laubacher, S.B. Dynes, Y. Zhao, Visualization of communication patterns in collaborative innovation networks-analysis of some w3c working groups, in *Proceedings of the Twelfth International Conference on Information and Knowledge Management* (ACM, 2003), pp. 56–60
7. Y.H. Kidane, P. Gloor, Correlating temporal communication patterns of the Eclipse open source community with performance and creativity. Comput. Math. Organ. Theory **13**(1), 17–27 (2007)
8. T. Laferrière, C. Hamel, M.D. Hamel, C. Perreault, S. Allaire, *Sustaining innovation in a networked classroom: The contribution of pre-service students*. Paper presented at the annual meeting of the American Educational Research Association (AERA), Washington, DC (2016)
9. M. Scardamalia, C. Bereiter, Knowledge building: theory, pedagogy, and technology, in *Cambridge Handbook of the Learning Sciences*, ed. by K. Sawyer (Cambridge University Press, New York, 2006), pp. 97–118
10. A. Toffler, *Powershift: Knowledge, Wealth and Violence at the Edge of 21st Century* (Bantam Books, New York, 1990)
11. UNESCO, *Towards Knowledge Societies* (Author, New York, NY, 2005)

Chapter 15
Fantastic Interfaces and Where to Regulate Them: Three Provocative Privacy Reflections on Truth, Deception and What Lies Between

Gianluigi M. Riva

Abstract Speech Interfaces represent a new interactive phenomenon, which entails massive personal data processing. The spectrum of legal issues that arises from this interaction impacts both user privacy and social relationships. This study addresses three potential issues or 'provocations' relating to speech interaction that illustrate the challenges and complexity of this socio-legal domain: (i) the potential for lying; (ii) the possibility of breaching the law; (iii) the ability to interpret an order. It deploys an in-depth analysis of the related legal consequences and implications with the scope to prompt discussion around these provocative issues. It first provides an overview of the correct hermeneutical approach to frame legal paradigms, highlighting the crucial legal aspects, conceptual approaches and interpretations to be considered when addressing the whole 'interactive artificial agents' (IAA) phenomenon. The study adopts the classical Civil Law system's methodology (qualitative/top-down analytical). The core of the study then focuses on the three provocations as connected by personal data processing. The goal is to provide a critical legal analysis of those interfaces that could impact the foundation of human socio-legal interrelations. By raising awareness of these controversial aspects, the work contributes to fostering further discussion about interdisciplinary privacy issues that stand at the intersection of Law, Social Sciences and HCI design, and that cross-pollinate each other.

Introduction

Interactive Artificial Agents no longer belong to the future or science fiction. Nevertheless, they still represent a new phenomenon, with a hype in the research field that matches their commercial success and spread in customers' homes and devices

G. M. Riva (✉)
School of Information and Communication Studies, University College Dublin, Belfield, D4, Dublin, Ireland
e-mail: gianluigi.riva@ucd.ie

Massachusetts Institute of Technology, Media Lab, Cambridge, MA, USA

© Springer Nature Switzerland AG 2020
A. Przegalinska et al. (eds.), *Digital Transformation of Collaboration*,
Springer Proceedings in Complexity,
https://doi.org/10.1007/978-3-030-48993-9_15

[1]. While they are also slowly entering the professional market by being adopted in offices [2], these interfaces are still the prerogative of a few producers: Alexa from Amazon, Siri from Apple and Google Assistant from Alphabet lead the market, and their features are to some extent equivalent. All these players aim to gain positioning that ensures the creation of an Internet of Things (IoT) connected ecosystem. The long-run goal with IAAs' is indeed to govern this IoT environment as the primary interface gate through which users interact with the whole smart ecosystem [3]. However, currently, IAAs are still stuck at the elementary level of interaction, performing simple tasks and providing pieces of information concerning users' questions [4]. Nonetheless, they represent a silent revolution that is slowly entering people's lives through a sort of Overton's Window,[1] based on the provision of comfort, services, task-management advantages and marketing strategies. The compensation users pay for these features is the personal data-commodification,[2] which implies that users enter contracts by adhesion[3] built around profiling activities as the cornerstone of how the service functions [6]. The scope—or the commercial excuse—is to 'improve the experience' of users.[4] This profiling activity represents the basis for legal and privacy concerns that may arise about IAAs [7], as both the users' personal data processing and related big data mining drive all the activities that these interfaces can perform. Consequently, data processing and mining are the activities that inform the whole analysis of the legal reflections tackled by this investigation.

This study aims to pose several provocations concerning legal, privacy, ethical and practical issues that arise from the interaction between users and interactive

---

[1] Which is defined as the 'range of ideas tolerated in public discourse, also known as the window of discourse' [5]. The concept states that an idea's political viability depends mainly on whether it falls within this range, rather than on politicians' individual preferences. Indeed, according to Overton, the window contains the range of policies that a politician can recommend without appearing too extreme to gain or keep public office in the current climate of public opinion. The range is formed by 'unthinkable', 'radical', 'acceptable', 'sensible', 'popular', 'policy', and describes the different stages of public perception.

[2] The relationship between users and service providers for the use of a service provided through an online platform (e.g. social media platforms) falls into the legal regime of contractual relationship. Despite service providers allege the service is provided for free (gratuitousness), this is not true: users pay the service (counter-compensate it) with their personal data, with their attention, and by accepting both to undergo advertisement, to be profiled and their profile to be sold to third parties. Therefore, for the Law, it is wrong to call them 'users', because they must more appropriately be considered 'clients' instead. For instance, note that the Facebook frontpage to access the platform after the Cambridge Analytica case has deleted the phrase 'it is for free and it always will be'. However, this misconception of the data processing relationship implies treating personal data as commodities (and properties), which is not the case. Under the GDPR and Civil Law, one does not own their personal data. Instead they have personhood rights over them (to license the data exploitation under certain conditions set by the regulatory framework). Privacy rights are fundamental rights and, as such, cannot (and must not) be monetised.

[3] Meaning those contracts in which the conditions cannot be bargained and essentially work as a 'take it or leave it' proposal.

[4] Although, the benefit of this improvement mostly privileges service providers' power of prediction and of shaping personalised advertisement.

artificial agents. The issues are approached by conducting a critical legal analysis of the phenomena, in light of the socio-legal effects that particular situations can entail. Indeed, as no specific regulation is in place for these phenomena, there is, therefore, the need to apply analogically[5] the general legal principles that govern the related theoretical paradigms. Thus, the method adopted is an empirical hermeneutic reasoning based on the legal syllogism technique, grounded on the classic critical legal approach and a legal qualitative/speculative analysis. The goal is to prompt discussion around these issues without providing definitive answers towards identifying key themes for further development. However, providing legal answers for a new phenomenon always requires that we bear in mind how the compass of regulation works: the Law[6] focuses on the effects, rather than on the actions; it often relies on presumptions,[7] discriminates among legal statuses; protects weaker parties; ensures accountability[8]; and adopts a neutral approach toward technology.[9]

Background

The world of speech interfaces is becoming crowded. Siri, Cortana, Alexa, Bixby and Google Assistant are the principal players. What these assistive interfaces certainly share is that they are non-embedded Artificial Intelligence (AI) agents, usually supported on mobile devices as interactive voice interfaces. They are based on in-cloud natural language processing (NLP) technology that allows two-way voice interaction between user and machine [8]. This involves many issues on both legal and ethical levels [9, 10] not least invasive personal data processing [11] and represents the first big step toward a full IoT ecosystem [12]. Their functionality entails constant data gathering/processing, machine-to-machine (M2M) interaction and continuous personal tracking with related legal, ethical and privacy implications [13].

However, despite the fame that IAAs are gaining among both customers, professionals and researchers, there is no unique name or categorisation for them.[10] For

---

[5]So-called '*Analogia Iuris*', i.e. the technique of applying, in absence of a specific regulatory framework, the legal regime provided for a similar phenomenon, e.g. what happened with airplanes at the beginning of the twentieth century, when were mutated navigation rules for regulating the phenomenon.

[6]When the term "Law" is capitalised, it refers to the whole Legal System. Instead, when it is not capitalised, it refer to a single law or norm.

[7]That is legal fictions: because it is not possible to measure, quantify, define or prove certain situations (acts, facts, conducts) the Law assumes that under certain circumstances there is a fixed outcome, called 'presumption'. The party who undergoes the effects of legal presumptions is usually admitted to prove the contrary.

[8]To provide stability of social relationships through enforceability.

[9]It regulates—through standardisation—the technical requirements that a technology must match in order to achieve the wanted effects.

[10]The literature refers to these devices indifferently as conversational agents, voice user interfaces, speech interfaces, digital or virtual assistant, smart speakers, and so on.

the sake of legal argumentations,[11] we refer to these interfaces as 'Interactive Artificial Agents' (IAA),[12] or 'speech interfaces', as per their general characteristics: they interact; they are human-made (artificial); act (have an agency with effects over the world and related socio-legal relationships); imply speech (both actively: vocal interaction; and passively: listening); and represent an interface through which both humans and machines can interact.

IAAs belong to the realm of NLP, namely a field at the intersection among Computer Science, Linguistic and Information. It concerns the linguistic interactions between human and computers, and it focuses on how programming computers to process natural language data [16]. NLP agents are not a novelty, and they were theorised since the '50s of the twentieth century [17]. Also, their practical application in bots dates to the '70s with the ELIZA project [18]. However, they gained momentum in the 2010s thanks to advances in Neural Networks and Deep Learning. Currently, Google Duplex represents the most progressed item in the field [19]. Nowadays, an estimation of 10% of world consumers own an IAA and they will soon reach the level of penetration that smartphones have [20].

While the literature around these technologies is flourishing across many disciplines, legal contributions are scarce and tend to focus on AI in general or specific AI-related aspects [21–23]. In turn, legal scholarship is beginning to address IAA privacy issues but mostly concerning their security aspects [7, 11, 13, 24–26]. On the other hand, the literature has also shown an interest in the trust aspect by focusing principally on the ethical level [27, 28]. Nevertheless, there is still a requirement to connect more fully different field reflections under a legal conceptual umbrella. To the author's knowledge, no investigation yet has tackled the connection between information management (data processing, interaction, filtering), interface design, Data Protection and the overall legal implications that derive from these. This study provides a holistic understanding of how concepts such as truth, filtering, interpretation and information control connect with privacy, as well as how they inform the relational paradigms between users and the interface. The red thread that underpins this investigation is tied to the legal effects that such concepts entail and that legislators urgently need to address.

Framing the Context

Since the 1930s, many science fiction novels, films and comic books have shaped our conceptual approach to both the future and to technological development. It is with

---

[11] Consider that, in Civil Law systems, the Law must be general and abstract, meaning that it should affect the most possible people and embrace the most possible situations.

[12] We avoid the use of 'intelligent' as the Law does not care about the intelligence level of an agent (be it natural or artificial), unless constituting incapability, because it focuses on actions and, overall, their effects. The quality (legal status) of the agent affects only its imputability. The point for the Law is not how much clever one is but if an agent can be considered as a legal subject instead of a legal object. This goes for robots (embodied AIs), as well as for animals [14, 15].

these lenses that we are unconsciously used to seeing the world. Nevertheless, using the correct paradigms is a fundamental activity for the Law, especially for regulating a phenomenon. For example, we might be inclined to perceive two different situations if we see a self-driving car travelling the streets without a pilot and a regular car driven by an android robot [29]. Actually, these represent the same phenomenon: an AI which drives a car. Therefore, in order not to confuse one with the other, we need to consider the difference between shape and use, as well as between product and service [30].

Similarly, decades of images which have depicted personal assistants as an individual's embodied-robot unit have corrupted this concept: i.e. as an isolated unit owned by the possessor, which only responds to the owner's orders as a personal butler, secretary, assistant or friend. The owner–robot relationship with the producer was imagined perhaps only in relation to replacement parts, whereas now it is a continuous real-time connection. The reality is, however, slightly different from the common perception: current technologies provide us with several forms of non-embedded assistant interfaces under the form of a service which is managed, trained and updated by the producer/service provider via the Internet. Users only own the 'shell' but have no real power over the 'ghost'. Thus, for legal analysis, the phenomenal paradigm must shift from a product to a service relationship. This service, in turn, is characterised by essential elements that render it subjected to the supplier's power and will, rather than the user's. Indeed, the service is far from being a continuous supply regulated by a previous unique agreement, such as for instance, an electricity supply.[13] On the contrary, it is an on-demand generic support (assisting interface) in which consumers only use the service 'for free', and only pay for the supporting device (the product). However, the contractual/data protection relationship between user and service provider is far from being without cost, as it relies on the user yielding non-necessary personal data.[14] Furthermore, users are obliged to see (or to ear) unwanted advertising,[15] they cannot negotiate the terms of the agreement and have no control about their personal profile sold to third parties. Therefore, the paradigmatic elements that inform this external (service providers') power in the user-interface relationship are (i) recurrent software and terms updates[16] which are outside users' control, (ii) constant personal data processing (profiling), as well as (iii) the related policies unilaterally provided by the supplier, (iv) a necessary connection via broadband, (v) the exposure to third-parties' application services and (vi) a potential conflict between users' and suppliers' interests.

---

[13]It is instead shaped as a licensee agreement of use, paired with privacy policies by adhesion.

[14]According to the GDPR the data processing must be informed by the principle of purpose, strictly linked with the principle of necessity and minimisation. They imply that the processing of personal data must be carried out for specific purposes only, as well as, the data processed must be quantitatively and qualitatively minimised to the extent of those solely necessary to provide the requested service.

[15]Users cannot avoid advertisement nor, usually, can opt-out from personalised ads.

[16]Usually dispatched together, so that users are forced to accept by adhesion both with no ability to opt-out from one or the other or single provisions.

We must also take into account that the relational paradigm shifts from physical usage to voice interaction (sequences of verbal contracts). All these paradigms have important impacts on which kind of regulation should apply to the relationship between users, interfaces and service suppliers. They also impact severely on the outcomes of the actions that the interfaces themselves perform, be these actions ordered by the users or performed towards third parties.

Three Legal Provocations for Many Practical Reflections

In light of these concerns, this study poses three questions which state provocative legal reflections. The scope of these questions is to foster discussion around potential issues that may plausibly arise from the usage of IAAs, by showing that there is no regulatory reference to frame these situations. It is not the goal of this study to provide answers to these questions. Indeed, answering the posed questions would necessitate—at least—a dedicated study focused on building the proper legal reasoning as a foundation for such answers. Indeed, in the absence of a regulatory framework or jurisprudence to guide the legal interpreter, the only way to provide answers is by adopting legal analogy and legal logic as guiding approaches.

Should IAAs Be Allowed to Lie to Users (and Vice Versa)?

The fact that an IAA system is capable of lying is not debated, but whether it should be allowed or not to do it, is a different question with many practical and legal implications. The current terms and conditions, service use conditions or privacy policies [31–34] of IAAs do not inform users if the system is allowed to lie or not. However, if one asks Alexa if it lies, it replies: "I'm not always right, but I would never intentionally lie" (other speech interfaces give similar responses)[17]. Although it is understandable from a contractual perspective that service providers shape their speech interfaces this way to avoid some liability, this can also have other implications [35]. Several studies claim that when the system cannot either lie or tell the truth (for whatever reason), it turns off[18] [36]. This occurrence seems another way to confirm, or at least to raise the doubt, that the system has turned off in order to avoid the dichotomy of both telling a lie or the truth. In any case, when it comes to complicated human–machine natural language relationships, the 'system cannot lie' oversimplification involves many issues. First, the avoidance

---

[17]For instance, Siri replies 'I'm not programmed to lie', while Google Assistant replies 'I'd never lie to you'.

[18]However, this is merely empirical. The terms and conditions do not cover this situation and there is no proof that the system turns off not to lie or to disclose the truth. However, it remains a fact that it does it every time for the same set of questions.

of lying does not imply perforce telling the truth. Many levels of communication are in place in a natural language interaction, including misinformation, rephrasing, omissions and bypassing answers. A pure and simple lie avoidance may represent one of those ethical settings that should be designed according to the Privacy by Design principles,[19] letting users decide in advance what the IAA should or should not do. Nevertheless, the true–false spectrum introduces new levels of complexity that are not addressable with a general case-study approach or automatic settings applicable to every situation as many nuances are in place.

First, the system can lie for particular apps, games or jokes, but this represents only an exception [37]. Secondly, applications are often third-parties' services and, as long as the API terms and conditions of IAAs do not mention the truth requirement, these apps might be designed with 'deceitful features' [38]. This may eventually generate confusion among users as well as issues of reliability and neutrality, which introduces uncertainty in socio-legal relationships.[20] Furthermore, the system might 'think' it is telling the truth, meaning it relies on incorrect sources, biases or merely non-updated state of the art, or even errors [39]. Conceptualising the truth-requirements implies speculating about the kind of 'truth'—or its degrees—that must be required for an IAA. Indeed, it can be argued that an IAA for a professional environment needs a technical/scientific truth, while in a social environment it needs to express accurately an acceptable truth. If a user asks the system to evaluate their weight based on biometric parameters, the system might respond with the precise measure or something such as 'you are fat'. Both can be true, but the second implies a socially unacceptable phrasing (with judgement), which in turn may involve tort law for insults. We may even speculate that the system should tell a lie to make users feel better and maintain the relationship, or not to harm people in sensitive psychological condition.

However, one can also perform deceit by bypassing answers framed with 'according to' or with contextual answers formally connected to the question but that substantially avoid it. For instance, if an IAA replies to an answer regarding personal data processing by providing the link to the privacy terms and conditions, it formally answers but substantially forces the user to read it through to find the answer.

Lying scenarios are far from being speculative and involve relevant legal issues, for instance, about children and third-party's private information disclosure, aside from tort law in the case of deceptive information. Indeed, when it comes to children (and minors in general), sometimes they may have to be protected from the truth, and not only with an omission or a lie. This means they should be protected by the existence of a specific sensitive topic in the first place as might also be the case with third-parties' personal information. The IAA should be able to understand the legal status and legal capacity of particular users and to interact with them accordingly, without disclosing sensitive or third-party information. On the other hand, third-party

---

[19]GDPR Article 25.

[20]Consider that the goal of legal systems is to provide a twofold tool for social stability (reliability of socio-legal relationships) which is represented by foreseeability and certainty.

information might relate to public personalities and so, non-disclosure could clash with the public interest in receiving public news. Further complications arise when the interaction occurs contemporarily with a plurality of individuals with different legal statuses (e.g. minors and adults). The system should be able to recognise the presence of minors and avoid certain pieces of information. An IAA should be able to provide 'white lies' to children to protect them, for example, in relation to the existence of Santa Claus. Furthermore, parents' and minors' right to privacy can overlap or be in conflict. The extent of a parent's right to access the minor's interface data is still under debate [40] as it increasingly conflicts with minors' privacy and depends on the age or the information that parents would like to access. However, if a third-party hosts children in an environment that includes an IAA, minors could go unprotected, and the practical capability for their parents to access their data would be hard to put in place.

As we can see, the IAA information-management approach plays a critical role in this issue. As speech interfaces process relevant personal information, they may be able to disclose this information to others in order not to lie. The case is still theoretical because the current IAA state of the art does not allow such a level of interaction. Nevertheless, it is worth considering that an individual should be protected from the disclosure of confidential information without their awareness, i.e. personal information. Therefore, if a jealous individual asks an IAA at which hour their partner came home the previous night, the system should not be allowed to disclose the information.[21] Or should be?[22] On the other hand, sharing personal information could be useful in a family environment, and such limitations undermine the usability and utility of the system. Again, allowing personalisation of the settings through Privacy by Design could be a step in the right direction.

In general, the issue concerning the system's capability for lying should be framed correctly, not only as the possibility to do so but also the intention to do it. Indeed, what we are truly asking is not if the IAAs lie or not, but if the underlying natural language algorithm can be trained to respond with a false answer on purpose. In terms of potential possibility, the answer would be reasonably affirmative. However, it does not seem to be the case, as IAAs are composed of both pre-recorded answers and NLP algorithms that appear not to be coded to do so [42]. Nevertheless, the issue is that this kind of decision is governed uniquely by service-providers. Users cannot opt to have the feature or gain control over it, even if these essential characteristics might impact on users' fundamental rights.

On the other hand, are users free to lie to speech interfaces and if not, should they be? It may seem obvious to answer affirmatively. Nonetheless, lying to (or through) an IAA may result in harm, which implies liability. Indeed, if we consider a complex interaction with many parties, this can easily happen, e.g. one tells the system that the gas bill is paid (while in truth it is not) and it confirms it when someone else asks for it, causing damages (debt protest). This kind of scenario is particularly true when

---

[21] Articles 5 and 32 of the General Data Protection Regulation (GDPR) [41].

[22] There is no discussion yet in the literature on the issue about knowing personal information of partners and the extent of this hypothetical right.

it comes to IAAs in work environments, in which the interface will interact with employees on behalf of the employer. In this case, one would be inclined to say that the employee should not lie to the system but, on the other hand, we can argue that it is their right to do so (or to omit). Indeed, this scenario involves not only fundamental personhood rights but also all the third-party legal relationships. Although it may be claimed that internal policies could regulate these cases, this would involve a shifting question: would policies that require people not to lie to speech interfaces be legitimate? More generally, there is also an open issue concerning the potential use of the lie. Admitting one can lie to the IAA, we should consider that this feature is recorded and can be used to profile the user as a liar and sell the information to third parties, impacting the user's external socio-legal sphere of relationships. Finally, the extent to which the lie can or cannot be used—and in which cases—against the liars themselves, remains debatable.

To What Extent Should IAAs Be Capable of Interpreting Orders?

When we interact with each other as humans, we take for granted many things about language that we do not even consider. For instance, when we ask someone to do something, e.g. 'go there', we take for granted that the recipient of the request will perform the action considering obstacles and the environment, as well as the whole non-verbal meaning that we often hide in requests ('go there' [by walking and wait for my next instruction]). Interaction with an IAA still does not work this way. It is the case of the Robot that walks on the table when asked to do so, which must be trained to stop when arriving at the table edge, in order not to fall [43]. While this is 'interpretation', it also means stressing the meaning of a concept, reformulating it according to the context or even by-passing an order. This banal reflection opens a wide breach in user-interface interactions concerning an IAA's capability to interpret in this sense (bypass) orders, requests, actions, situations and even regulations. There are many implications from this. Interpretation is undoubtedly useful, but in our daily experience, it can also result in mistakes and harmful consequences (or lies we are unaware of). As humans, we have a system (the Law) to allocate responsibility, but this still does not address artificial agents [44]. Once IAAs will be able to perform such a task, we will have to solve the accountability issue for harms derived by the wrong (or right) interpretation. This implies a subsequent issue: society accepts legal punishment as it inflicts a retribution cost suffered by the guilty [45]. Damaged parties are not merely compensated by the monetary punishment but also by the awareness that the guilt will suffer the (emotional and material) harmful consequences of paying it. Even if we design legal personhood for artificial agents [46], we will always lack that kind of retribution, which eventually could undermine the scope of account- ability allocation. Furthermore, on another level, this relates to our antithetic desire to demand the artificial agent to act automatically and flawlessly, but also ethically and appropriately considering both emotional and moral contexts. This need for retribution, may create many 'false' (legal) issues, such as the trolley dilemma[23] [47], in which non-legal scholars neglect to consider legal solutions such as the state of necessity[24] and the regime for lawful justifications.[25]

Nevertheless, another critical issue is how to determine the criteria to embed this interpretative ability in IAAs and, above all, who should decide. This involves the 'black box issue' and the necessity for explainability and transparency [48]. The speech interface should always be able to explain why it interpreted something in one way instead of another and to keep track of the process. However, we are black boxes too, and the Law already provides for us thanks to legal fictions, objective accountability, negligence rules and justifications. Thus, the IAA black box interpreter falls into the GDPR regime for explainability.[26] Though, Article 22 on Automated individual decision-making may represent a barrier for automated interpretation. Indeed, the said provision requires the so-called human-in-the-loop for automated data processing that produces legal effects concerning data subjects or, that significantly affects them. The provision refers to both 'decisions', 'legal effects' or just significant implications that concern data subjects. This point is crucial because it creates a conflict, as the IAAs able to interpret users' orders could not lawfully function unless mediated every time by a human who eventually ratifies the outcome.[27] In other circumstances, this could even represent a socio-ethical dilemma: in a medical environment in which a professional IAA (e.g. Watson Health [51]) would assist/advise doctors, the latter always should have, reasonably,[28] the last word over the machine. Notwithstanding, this recognition of the power of evidence mined through machine-learning might result in the ratification of the IAA's decision in order to avoid any potential liability or proof against the machine.

However, one further crucial privacy issue comes from interpretation, as interpreting also means the ability to discriminate among different individuals [49]. Although discrimination—which also implies profiling—can be useful in some situations, this aspect reflects how much power an artificial agent (and its service provider) would gain in this scenario. This is why limitations such as Article 22 are in place, although they may practically undermine the (future) functioning of an actual IAA.

---

[23]Often (legally) misused to approach self-driving car scenarios, their liability consequences and the related impact on AI systems programming.

[24]It is not punishable whoever committed the fact for being forced to do so by the need to save himself or others from the current danger of serious harm.

[25]The so-called 'excitement causes' which are established by the law and identify particular situations whose occurrence makes it lawful to commit a crime. For instance, self-defence.

[26]Article 22.

[27]However, consider that the WP29 guidelines [49, 50] underline how this should not be the case and, instead, the human intervention in the decision-making process should be substantial.

[28]And according to GDPR Article 22.

Should IAAs Breach the Law if Requested (and Allow It)?

To correctly answer this question, we must define what we mean by 'the law'. The Law, indeed, is a body of different hierarchical sources that have different aims and degrees of importance. Hence, there is a difference between violating a fundamental right or an administrative law. On the other hand, violating freedom of speech might be less (materially) harmful than breaking a red traffic light. Furthermore, there might be cases in which violating the law can be admissible, for instance, for saving a life under the state of necessity. We also must consider that an IAA may be used to harm someone else. In this case, IAAs should be able to interpret the situation for assessing whether or not it is reasonable to break a legal (or social) provision, implying all the issues already addressed. In turn, we may reverse the question to understand if the IAA should or should not allow the user to breach the law. Here, we have the same kind of conceptual considerations on paradigms already performed. Indeed, if we instead ask whether laptop producers should allow users to exploit the device to breach the law, we might see the situation from a different perspective. Here the paradigm describes the IAA as a device, meaning a tool, managed according to a contractual relationship with the user. Current speech interfaces already prevent users from accessing potentially harmful information, which could be used to commit crimes. If we ask the system how to make a bomb, the system does not provide the answer. This feature works like a pre-crime preventive filter, which, however, limits the individual freedom of accessing knowledge[29] [49].

In general, this question opens an intriguing legal–ethical dilemma as well as an essential issue for the Philosophy of the Law: is breaching the law a non-written fundamental right? Indeed, the law provides rules and punishment for their breach so that an individual has the choice between following the rule or accepting the consequences and risks of its breach[30] [50], which is an essential expression of the right of self-determination. However, it does not mean that one should be able to kill someone else only because they choose not to follow the law, but still, this conceptual choice has fundamental rights implications on preventive policies. For instance, we can easily kill someone with a knife and yet we can find knives at a restaurant. In some places, the law prescribes knives to be rounded for this reason, and this is how the Law usually approaches such issues. Eventually, this issue is connected to the ability of future IAAs and smart IoT environments to monitor users and the social tendency to claim we should prevent crimes in this passive-aggressive way. However, consider that if we say that we should prevent every breach of the law, we imply that constant monitoring and surveillance of users is not only fair but should be lawful and even implemented.

---

[29]And furthermore, it is against the general penal principles of Civil Law systems: one cannot be charged for the mere intention to commit a crime and, therefore, cannot be prevented to express that intention. For the Law, it is the action that is relevant.

[30]Which, in both Economy and Criminology, falls into the rational-choice Theory, which posits that humans are reasoning actors who weigh means and ends, costs and benefits, through utilitarian approaches in order to make a rational choice [52, 53].

Conclusion

This study has outlined some characteristics of interactive artificial agents that are relevant to current discussions around privacy and has framed a perspective for understanding some legal reflections. The contribution has focused on three provocative reflections about (i) the opportunity for IAAs to lie and related legal implications; (ii) their capability to interpret orders and what this entails in relation to socio-legal impacts; and (iii) their potential to breach the law and the consequences of doing so. The study has investigated the common thread of data processing and privacy issues that underpin these questions and how they interrelate in a complicated scenario of legal and social relationships. It provided initial light over reflections that the literature still does not consider and opened the path for further investigations at the intersection with Privacy, HCI and Ethics regarding IAAs. Future research may want to expand the investigation around these issues with in-depth analysis of connected Court decisions or a comparative study in relation to different jurisdictions.

Acknowledgements This investigation has been carried out thanks to Fulbright-Shuman Grant scheme for Visiting Scholars' Research Projects. The author would like to thank Dr. Marguerite Barry, which has no responsibility for the content, for her precious suggestions.

References

1. S. Perez, Over a quarter of US adults now own a smart speaker, typically an Amazon Echo (2019), https://techcrunch.com/2019/03/08/over-a-quarter-of-u-s-adults-now-own-a-smart-speaker-typically-an-amazon-echo/. Last accessed 7 Dec 2019
2. S. Forsdick, Amazon launches Alexa for business, but will office smart speakers take off? (2019), https://www.ns-businesshub.com/technology/amazon-launches-alexa-forbus iness/. Last accessed 7 Dec 2019
3. K. Noda, Google Home: smart speaker as environmental control unit. Disabi. Rehab. Assis. Technol. (2017). https://doi.org/10.1080/17483107.2017.1369589
4. L. Clark, N. Pantini et al., What makes a good conversation? Challenges in designing truly conversational agents, in *2019 Conference on Human Factors in Computing Systems (CHI 2019).* arXiv:1901.06525 [cs.HC] (2019)
5. J.P. Overton, https://www.mackinac.org/OvertonWindow. Last accessed 7 Dec 2019
6. T. Wu, Blind spot: the attention economy and the law, Columbia Law School (2017), Scholarship archive. https://scholarship.law.columbia.edu/cgi/viewcontent.cgi?article=3030& context=faculty_scholarship. Last accessed 27 Dec 2019
7. H. Chung, S. Lee, Intelligent virtual assistant knows your life (2018), arXiv:1803.00466 [cs.CY]
8. S. Sekine, H. Wakahara, Natural language processing device, method, and program. EP2653981A1 European Patent Office (2011)
9. N. Bostrom, E. Yudkowsky, The ethics of artificial intelligence, in *Cambridge Handbook of Artificial Intelligence*, ed. by W. Ramsey, K. Frankish (Cambridge University Press, 2011)
10. F. Raso, H. Hilligoss et al., Artificial intelligence & human rights: opportunities & risks (2018), https://cyber.harvard.edu/publication/2018/artificial-intelligence-human-rights. Last accessed 7 Dec 2019
11. C. Chhertri, V.G. Motti, Eliciting privacy concerns for smart home devices from a user centred perspective, in *iConference 2019*, ed. by N.G. Taylor et al. (LNCS 11420, 2019), pp. 91–101

12. P. Cohen, A. Cheyer et al., On the future of personal assistants, in *Proceedings of the 2016 CHI Conference Extended Abstracts on Human Factors in Computing Systems* (ACM, 2016), pp. 1032–1037
13. H. Schnädelbach, D. Kirk (eds.), *People, Personal Data and the Built Environment* (Springer, 2019)
14. E. Stradella, P. Salvini et al., Robot companions as case-scenario for assessing the "subjectivity" of autonomous agents, in *CEUR Workshop Proceedings on Some Philosophical and Legal Remarks*. (2012), http://ceur-ws.org/Vol-885/paper4.pdf. Last accessed 27 Dec 2019
15. E. Schaerer, R. Kelley et al., Robots as animals: a framework for liability and responsibility in human-robot interactions, in *18th IEEE International Symposium on Robot and Human Interactive Communication Toyama* (Japan, 2009)
16. Y. Goldberg, A primer on neural network models for natural language processing. J. AI. Res. **57**, 345–420 (2016)
17. A. Turing, Computing Machinery and Intelligence. Mind **LIX**(236), 433–460 (1950)
18. J. Weizenbaum, *Computer Power and Human Reason: From Judgment to Calculation* (W.H. Freeman and Company, 1976), pp. 2, 3, 6, 182, 189
19. Google Duplex. Google blog. https://ai.googleblog.com/2018/05/duplex-ai-system-for-natural-conversation.html. Last accessed 7 Dec 2019
20. R. De Renesse, Virtual digital assistants to overtake world population by 2021 (2017), https://ovum.informa.com/resources/product-content/virtual-digital-assistants-to-overtake-world-population-by-2021. Last accessed 7 Dec 2019
21. R. Leenes, F. Lucivero, Laws on robots, laws by robots, laws in robots: regulating robot behaviour by design. Law Innova. Technol. **6**(2), 193–220 (2014)
22. F. Andrade, P. Novais et al., Contracting agents: legal personality and representation. Artifi. Intell. Law **15**(4), 357–373 (2007)
23. W. Wallach, From robots to techno Sapiens: ethics, law and public policy in the development of robotics and neurotechnologies. Law Innov. Technol. **3**(2), 185–207 (2011)
24. N. Abid, K. Ramokapane, J.M. Such, more than a smart speakers: security and privacy perceptions of smart home personal assistants, in *Proceedings of the 15th Symposium on Usable Privacy and Security* (USA, 2019), pp. 451–466
25. M.M. Losavio, K.P. Chow et al., The internet of things and the smart city: legal challenges with digital forensic, privacy and security. Secur. Privacy **1**(23), 1–11 (2018)
26. P. Cheng, I.E. Bagci et al., Smart speaker privacy control—acoustic tagging for personal voice assistants, in *IEEE Workshop on the Internet of Safe Things* (SafeThings 2019) (2019)
27. Y. Liao, J. Vitak et al., Understanding the role of privacy and trust in intelligent personal assistant adoption, in *iConference 2019*, ed. by N.G. Taylor et al. (LNCS 11420, 2019), pp. 102–113
28. Z. Rutkay, C. Pelachuad (eds.), *From Brows to Trust Evaluating Embodied Conversational Agents* (Kluwer Academic Publisher, 2004)
29. N.M. Richards, D.S. Smart, How should the law think about robots?, in *Robot Law*, ed. by A.M. Froomkin, I. Kerr, R. Calo (Edward Elgar Publishing, 2016)
30. M. Alovisio, C. Blengino et al., *The Law of Service Robots*, ed. by C. Artusio, M.A. Senor. Report (NEXA Center for Internet and Society, 2015)
31. Alexa and Alexa device terms, https://www.amazon.com/gp/help/customer/display.html?nodeId=201566380. Last accessed 7 Dec 2019
32. Google Assistant terms of service, https://developers.google.com/assistant/console/policies/terms-of-service. Last accessed 7 Dec 2019
33. Apple software license agreements, https://www.apple.com/legal/sla/. Last accessed 7 Dec 2019
34. Microsoft cortana and privacy, https://support.microsoft.com/en-us/help/4468233/cortana-and-privacy-microsoft-privacy. Last accessed 7 Dec 2019
35. S. Tiribelli, Nuovi Media e Bellezza. La narrazione estetica del sé tra potenziamento e falsificazione. In Etica e Bellezza, ed. by S. Achella, F. Miano. SIFM 3, 263–272 (2019)
36. K. Cox, Why Amazon's Alexa can't tell you if it's connected to the CIA (2019), https://www.consumerreports.org/consumerist/why-amazons-alexa-cant-tell-you-if-its-connected-to-the-cia/. Last accessed 7 Dec 2019

37. Amazon skills. https://www.amazon.com/Cloudlands-Dev-One-Lie/dp/B07GLN2D2L. Last accessed 7 Dec 2019
38. Google Assistant API terms of services, https://developers.google.com/assistant/sdk/terms-of-service. Last accessed 7 Dec 2019
39. Apple discussion, https://discussions.apple.com/thread/250073486. Last accessed 7 Dec 2019
40. L. Bolognini, C. Bistolfi, L'età del consenso digitale. Privacy e minori on line, riflessioni sugli impatti dell'art. 8 del Regolamento 2016/679(UE). CNAC report (2017)
41. Regulation (EU) 2016/679 of the European Parliament and of the Council of 27 April 2016 on the protection of natural persons with regard to the processing of personal data and on the free movement of such data and repealing Directive 95/46/EC (General Data Protection Regulation)
42. D.A. Moses, M.K. Leonard, Real-time decoding of question-and-answer speech dialogue using human cortical activity. Nat. Commun. **10**, 3096 (2019). https://doi.org/10.1038/s41467-019-10994-4
43. Robot refuses to follow orders given by a human, https://www.dailymail.co.uk/video/news/video-1232192/Robot-follows-order-walk-table-trusting-caught.html. Last accessed 7 Dec 2019
44. A. Bertonilin, Robots and liability: justifying a change in perspective, in *Rethinking Liability in Science and Technology*, ed. by F. Battaglia, M. Nikil et al. (Pisa University Press, 2014)
45. J. Danaher, Robots, law and the retribution gap. Ethics Inf. Technol. **18**, 299–309 (2016)
46. R. Whiters, The EU is trying to decide whether to grant robots personhood (2018), https://slate.com/technology/2018/04/the-eu-is-trying-to-decide-whether-to-grant-robots-personhood.html. Last accessed 7 Dec 2019
47. J.J. Thomson, Killing, letting die, and the trolley problem. Monist **59**, 204–17 (1976)
48. S. Wachter, B. Mittlestadt, L. Floridi, Transparent, explainable, and accountable AI for robotics. Sci. Robot. **2**, 6080 (2017)
49. G.M. Riva, Net Neutrality matters: privacy antibodies for information monopolies and mass profiling. Publicum **5**, 2 (2019), https://doi.org/10.12957/publicum.2019.47199. https://www.e-publicacoes.uerj.br/index.php/publicum/article/view/47199
50. A. Ravà, *Il diritto come norma tecnica* (Cagliari, Dessì, 1911)
51. PYMINTS, IBM gives Alexa another rival with Watson Assistant (2018), https://www.pymnts.com/news/artificial-intelligence/2018/ibm-alexa-watson-assistant-ai-smart-speaker/. Last accessed 27 Dec 2019
52. N. Garoupa, Economic theory of criminal behavior, in *Encyclopedia of Criminology and Criminal Justice*, ed. by G. Bruinsma, D. Weisurd, Chapt. 327 (2014)
53. R.V. Clarke, Situational crime prevention, in *Building a Safer Society: Strategic Approaches to Crime Prevention*, ed. by M. Tonry, D. Farrington (The University of Chicago Press, Chicago, 1995). ISBN 0-226-80824-6

Part IV
Interdisciplinary Methods

Chapter 16
A Structured Approach to GDPR Compliance

Antonio Capodieci and Luca Mainetti

Abstract The European General Data Protection Regulation (GDPR, EU 2016/679), adopted by the European Parliament has profoundly changed the legislative approach to the protection of personal data by the European Union. The GDPR provisions require organizations to make deep changes. Organizations have to shift from an approach based on the adoption of minimum-security measures, provided by the EU Directive of 1994, to a proactive approach based on accountability. Organizations that manage personal data of EU citizens have to adopt systems of verification and continuous improvement and adopt principles such as privacy by design and privacy by default. The rule of "privacy by design" calls for privacy to be taken into account throughout the whole engineering process. A key point is the methods for checking compliance with GDPR. This paper proposes a structured approach based on business process modelling, to support compliance with the GDPR. We have identified an approach that has to identify the most important key points for GDPR compliance.

Introduction

On 24 October 1995, the European Data Protection Directive (officially: Directive 95/46/EC on the protection of individuals with regard to the processing of personal data and on the free movement of such data) was created as an essential element of EU privacy and human rights law. The directive required EU member states to implement the corresponding provisions in national law by 24 October 1998.

The Charter Of Fundamental Rights Of The European Union, approved in 2007, (2007/C 303/01) says in Article 8—Protection of personal data that "Everyone has the right to the protection of personal data concerning him or her." and "Such data

A. Capodieci (✉) · L. Mainetti
Università del Salento, Lecce, Italy
e-mail: antonio.capodieci@unisalento.it

L. Mainetti
e-mail: luca.mainetti@unisalento.it

© Springer Nature Switzerland AG 2020　　　　　　　　　　　　　　　　233
A. Przegalinska et al. (eds.), *Digital Transformation of Collaboration*,
Springer Proceedings in Complexity,
https://doi.org/10.1007/978-3-030-48993-9_16

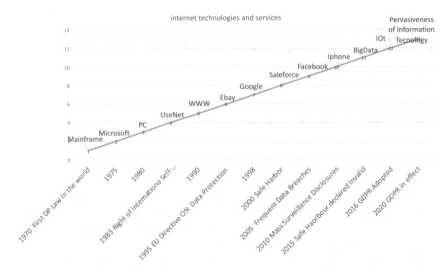

Fig. 16.1 Internet services and regulations evolution

must be processed fairly for specified purposes and on the basis of the consent of the person concerned or some other legitimate basis laid down by law. Everyone has the right of access to data which has been collected concerning him or her, and the right to have it rectified." Subsection 3 says: "Compliance with these rules shall be subject to control by an independent authority."

As we can see Fig. 16.1 [1], the panorama of "internet technologies and services" has completely changed since 1994 and the normative prescription became inadequate to protect personal data.

Finally, starting from 2016 the European Union, from principles enshrined in the Charter of Fundamental Rights, has completely revolutionized the regulatory framework regarding the protection of personal data. Several legislative provisions (regulations and directives) have been issued and others are in the process of being issued.

The EU has established a single privacy regulatory reference for all member states to give individuals more rights on their personal data. It is equally important that the standards apply to all companies that process data of European citizens, regardless of the place in which this data is processed.

The most well-known provision is the General Data Protection Regulation, better known as the GDPR.

The new important elements introduced by the GDPR are the self-assessment of risk to lose personal data and the definition of the strategy to reduce this risk.

In this context, the organizations collecting and processing personal data must be explicit about their motivation for data collection, who has access to the data, and how, when and how often the data will be used.

To be GDPR compliant, it is fundamental to know well all the activities related to personal data and all the employees that manage personal data.

A self-evaluation necessarily includes the understanding of all the business processes that are implemented and requires to identify who manages that personal data.

In the context of Computer Science, there are a variety of modelling languages, methodologies and tools focused on the concept of "process". For example DECLARE [2], DCR Graphs [3], State Charts [4], UML [5–8], GSM [9], CMMN [10] and Business Process Model and Notation (BPMN) [11]. All of these methodologies and tools were developed prior to the introduction of the GDPR. Therefore, during the process of analysis they do not require collecting all the data necessary to ensure compliance with GDPR.

Business Process Model Notation (BPMN) provides a standard and easy-to-read way to define and analyse business processes. BPMN reduces the distance between a process model and its implementation [12]. In the context of the Data Protection framework law, we believe that BPMN could be very useful and interesting.

The GDPR

The GDPR standardizes legislation for the management of personal data throughout the European Union.

Article 25 _Data protection by design and by default_ requires that data controllers focus on the protection of personal data in both the planning and organization of services, and in the stage of modelling of IT systems.

Subsection 1 says _"Taking into account the state of the art [...] the controller shall, both at the time of the determination of the means for processing and at the time of the processing itself, implement appropriate technical and organisational measures, [...], which are designed to implement data-protection principles, [...], in an effective manner and to integrate the necessary safeguards into the processing in order to meet the requirements of this Regulation and protect the rights of data subjects."_ In Subsection 2 it is required that _the controller shall implement appropriate technical and organisational measures for ensuring that, by default, only personal data which are necessary for each specific purpose of the processing are processed. [...]"_.

The records of processing activities [9, Article 30] are the main elements in the accountability of the owner, as they are useful in the recognition and evaluation of the treatments carried out, and also in the risk analysis and proper planning of treatments. The register must contain at least the following information: (i) the name and contact details of the data controller; (ii) the purposes of the processing, distinguished by types of treatment; (iii) a description of the categories of data subjects (e.g. customers, suppliers, employees) and the categories of personal data (e.g. personal data, health data); (iv) the categories of recipients (even by category only) to whom the personal data have been or will be communicated; (v) the latest deadlines for the cancellation of the different categories of data; (vi) a general description of the technical and organizational security measures referred to in Article 32.

In general, GDPR are statements about how an organization collects, processes and more generally manages the personal data of individuals.

Research Question

One of the most important requirements related to the GDPR is to explicitly describe all the activities that manage personal data and to formally define the complete life cycle of such data.

The problem is that there aren't methodologies and tools that can accomplish this goal. When modelling data, software engineering does not explicitly describe if a record is "personal data" and does not model the retention time of this data (i.e. how long the processor should keep a record of this data). The most popular software modelling tools don't pay attention to the life cycle of the data, and it's very difficult to know the processes and actors that manage personal data [12, 13].

BPMN is focused on the business process. The authors showed in [14–16] that a methodology, based on business process management analysis, could be very suitable when adapting business practices to emerging organizational forms.

In our work, we propose that a model based on BPMN can provide an appropriate basis to support the Privacy-by-Design approach.

Moreover BPMN could also be convenient to define the records of processing activities that include all the information required by the GDPR and can provide a solid support for auditing and compliance with GDPR, since BPMN is oriented to support the business analysis.

Related Work

The state of the art presents several studies where BPMN meets both security and privacy aspects. In [17], privacy concerns are captured by annotating the BPMN model. Brucker [18] extends BPMN with access control, the separation of duty, the binding of duty and the need to understand principles.

In [19], BPMN is extended with information assurance and security modelling capabilities. Altuhhov in [20], aligned BPMN with the domain model of security risk management. In [21], privacy-enhancing technologies (PETs) are applied to enforce privacy requirements and support the analysis of private data leakage. A query language for representing security policies and a query engine that enables checking are presented in Salnitri et al. [22].

Moreover, some works involve the definition of extensions of BPMN to meet cybersecurity requirements [23, 24]. In [25], Maines and colleagues investigate an approach to modelling security. The authors used BPMN choreography to model messages exchanged and identity contract negotiation.

Recent contributions, such as [15, 17, 26, 27] present specific BPMN security extensions in a healthcare context. The authors introduce security elements for BPMN in order to evaluate the trustworthiness of participants based on a rating of enterprise assets, and to express security intentions such as confidentiality or integrity on an abstract level [28].

Nowadays, a limited number of works have now studied the correlation between the GDPR and process management. In our work, we propose to extend BPMN with meta-information for each element of BPMN that classifies the element in the context of the GDPR.

There are a few studies that deal with GDPR, though from different points of view. The authors of [29] present a legal ontology for the GDPR that aims to provide a model of legal knowledge regarding the privacy agents, data types, types of processing operations and the rights and obligations involved. This work is very interesting, and we will use it as a basis for our work, as described above.

The authors of [30] propose an approach that identifies a "purpose" (purpose for data management as defined in GDPR) with a business process, and shows how to use a formal models of a process to derive privacy policies. Although this approach is also very interesting, in our opinion it is not able to identify all the information necessary to support GDPR compliance.

Finally, to capture security requirements within business process modelling, it is useful to have a notation that is supported by a set of graphical concepts, allowing us to represent the security semantics [27].

Our Solution

In this paper, we propose a structured approach based on the analysis of business processes, which allows the precise extraction of the records of processing activities with the necessary attributes. The proposed method also allows us to identify all data processors.

Extension of BPMN

Our approach is based on an extension of the BPMN, already published in [31].

We defined a set of meta-information for each element of the BPMN design that classifies the element in the context of the GDPR.

For each pool/process, our method indicates the following:

(i)   whether the process deals with personal data;
(ii)   the legal basis that authorizes its execution;
(iii) the period of time for which the data is stored. Each activity is classified as to whether it concerns personal data and the type of data processed.

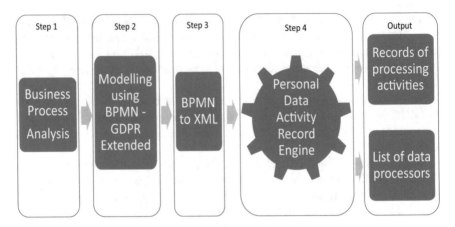

Fig. 16.2 BPMN vacation request with GDPR annotation [31]

In order to avoid creating custom BPMN notation extensions, the "tagged values" field was used. Based on the core definition of the BPMN, the appearance and specification of certain elements and connectors were defined by tagged values (Fig. 16.2).

The following tags were inserted in the pool element:

- GDPR: IsPersonalDataProcessing: a Boolean value (Yes/No) indicates whether the process involves personal data.
- GDPR: LegalBasis: contains references to the motivations for the execution of the process.
- GDPR: Duration: the period of time for which storage is expected.
- The following tags were inserted into the task element:
- GDPR: PersonalData: a Boolean Value (Yes/No) indicates whether the activity involves personal data.
- GDPR: TypeOfPersonalData: indicates the type of personal data processed (personal data, judicial, health data, political and religious opinions, biometric).

Proposed Approach: BPMN-GDPR-ENHANCED

As we can see in Fig. 16.3, our approach is based on five steps:

Step 1—Business Process Analysis: The science has demonstrated the validity of Business Process Analysis for understanding how organizations manage information and data with the necessary level of detail. Here we propose that the application of BPA can be used to explain and identify the processes that manage personal data. BPMN is a consolidated tool to model the Business Process.

Step 2—Modelling using BPMN-GDPR Extended: To discover how an organization manages personal data and to define the processor of such types of data, and

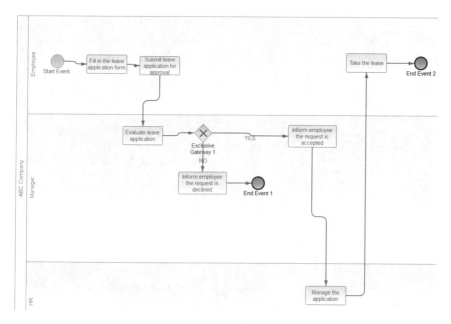

Fig. 16.3 Proposed approach: BPMN-GDPR-ENHANCED

to model activity (process) and actor (data processor) with the necessary information required for GDPR we propose to adopt an extension of BPMN (see the previous paragraph).

Step 3—BPMN to XML: In this stage, we export the BPMN models of our organization, identified in previous steps, in XML format. This allows us to have a machine-readable model of the Business Process and Actors.

Step 4—Personal Data Activity Record Engine (PDARC): The Personal Data Activity Record Engine is a tool used to query the XML model, from Step 2. PDARC is also able to extract the entire process and all the actors that manage personal data.

This is possible due to a classification based on the tagged values of the BPMN elements. The xml format of the BPMN diagram can be queried and a list of all the processes that have the tag "GDPR: PersonalDataProcessing" as equal to true can be extracted.

The output of this step is two elements fundamental to demonstrating compliance with GDPR: (1) Records of processing activities and (2) List of data processors.

Implementation

To implement our approach, we decided to extend a BPMN 2.0 Modeler tool. BPMN does not have a uniform implementation. Although it is defined as a standard.

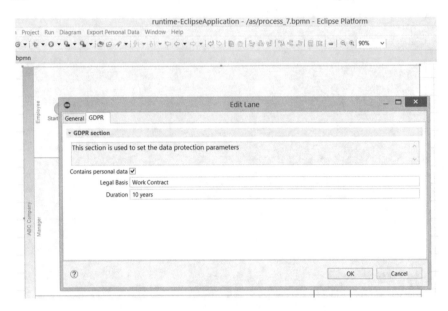

Fig. 16.4 The interface of the extension plugin

For our project we selected the Eclipse BPMN2 Modeler,[1] which is an Eclipse Plugin that implements the most important BPMN features. The foundation of the BPMN2 Modeler is the Eclipse BPMN 2.0 EMF meta-model, which is part of the Model Development Tools (MDT) project.

BPMN2 Modeler uses the Eclipse Plugin Architecture and provides several extension points for customizing the editor's appearance and behaviours.

We are developing an add-on of the BPMN2 Modeler that allows the modeller to add GDPR TAG to the BPMN process element.

In Figs. 16.2 and 16.4, we show an application of our approach to the process of a vacation request. While not necessarily a real business process, this example only aims to show the properties of our approach.

We can observe that the vacation request process involves the records of processing activities and is related to the management of personal data.

After the pool, we analysed the tasks in each lane.

For each task, we defined whether it requires the management of personal data (using the tag value "GDPR:PersonalData"), and what type of personal data is managed (using the tag value "GDPR:TypeOfPersonalData").

Figure 16.4 shows the interface of the extension plugin in operation and a sample of tagging activity of the Pool adding the value of Legal Basis and The Retention time (Duration).

---

[1] https://www.eclipse.org/bpmn2-modeler/.

```xml
<?xml version="1.0" encoding="UTF-8"?>
<!-- origin at X=0.0 Y=0.0 -->
<bpmn2:definitions xmlns:xsi="http://www.w3.org/2001/XMLSchema-instance" xmlns:bpmn2="http://www.omg.org/spec/BPMN/20100524/MODEL" xmlns:bpmndi="http://www.omg.org,
  <bpmn2:process id="process_7" name="Default Process" isExecutable="false">
    <bpmn2:laneSet id="LaneSet_1" name="Lane Set 1">
      <bpmn2:lane id="Lane_1" ext:IsPersonalDataProcessing="true" ext:Duration="10 years" ext:LegalBasis="Work Contract" name="ABC Company">
        <bpmn2:childLaneSet xsi:type="bpmn2:LaneSet" id="LaneSet_2" name="Lane Set 2">
          <bpmn2:lane id="Lane_2" ext:IsPersonalDataProcessing="false" ext:Duration="" ext:LegalBasis="" name="Employee">
            <bpmn2:flowNodeRef>StartEvent_2</bpmn2:flowNodeRef>
            <bpmn2:flowNodeRef>Task_2</bpmn2:flowNodeRef>
            <bpmn2:flowNodeRef>Task_3</bpmn2:flowNodeRef>
            <bpmn2:flowNodeRef>Task_8</bpmn2:flowNodeRef>
            <bpmn2:flowNodeRef>EndEvent_3</bpmn2:flowNodeRef>
          </bpmn2:lane>
          <bpmn2:lane id="Lane_3" ext:IsPersonalDataProcessing="false" ext:Duration="" ext:LegalBasis="" name="HR">
            <bpmn2:flowNodeRef>Task_7</bpmn2:flowNodeRef>
          </bpmn2:lane>
          <bpmn2:lane id="Lane_4" ext:IsPersonalDataProcessing="false" ext:Duration="" ext:LegalBasis="" name="Manager">
            <bpmn2:flowNodeRef>ExclusiveGateway_1</bpmn2:flowNodeRef>
            <bpmn2:flowNodeRef>Task_5</bpmn2:flowNodeRef>
            <bpmn2:flowNodeRef>EndEvent_2</bpmn2:flowNodeRef>
            <bpmn2:flowNodeRef>Task_6</bpmn2:flowNodeRef>
          </bpmn2:lane>
        </bpmn2:childLaneSet>
      </bpmn2:lane>
    </bpmn2:laneSet>
    <bpmn2:startEvent id="StartEvent_2" name="Start Event ">
      <bpmn2:outgoing>SequenceFlow_2</bpmn2:outgoing>
    </bpmn2:startEvent>
    <bpmn2:task id="Task_2" ext:IsPersonalData="true" ext:PersonalDataType="Not specified" name="Fill in the leave application form">
      <bpmn2:incoming>SequenceFlow_2</bpmn2:incoming>
      <bpmn2:outgoing>SequenceFlow_3</bpmn2:outgoing>
    </bpmn2:task>
    <bpmn2:task id="Task_3" ext:IsPersonalData="false" ext:PersonalDataType="No personal data" name="Submit leave application for approval">
      <bpmn2:incoming>SequenceFlow_3</bpmn2:incoming>
      <bpmn2:outgoing>SequenceFlow_4</bpmn2:outgoing>
    </bpmn2:task>
```

Fig. 16.5 BPMN-XML SCHEMA EXTENDED

As we can see in Fig. 16.5 our add-on adds special GDPR property to the XML schema of our BPMN MODEL. Querying the xml schemas we can extract only the element that has property ext:IsPersonalDataProcessing equal "true".

<bpmn2:process   id="process_7"   name="Default   Process"   isExecutable="false"><bpmn2:laneSet id="LaneSet_1" name="Lane Set 1">

<bpmn2:lane   id="Lane_1"   ext:IsPersonalDataProcessing="true" ext:Duration="10 years" ext:LegalBasis="Work Contract" name="ABC Company">

Based on XML schema, as shown before, our add-on implements the module the **Personal Data Activity Record Engine** PDARC that allows users to export the "Records of Processing Activities" and the "List of Data Processors" in different "formats", for example, Word, EXCEL, XML, CVS. Some organizations could adopt specific applications to manage activities related to GDPR, then could be useful to import directly in this system the "Records of processing activities" and the "List of Data Processors" coming from PDARC. At this moment, PDARC is still under development.

As mentioned earlier, we believe this approach allows us to determine whether a process involves records of processing activities, with the necessary information to file the records, as well as to define the data processors.

Conclusion and Next Steps

The proposed structured approach starts from the classical business process analysis integrated with the elicitation of information prescript from GDPR. From a properly noted BPMN diagram, we can extract the necessary information for the related records of processing activities (their legal basis, the duration of conservation, etc.). Similarly, it is possible to identify all actors who are data processors. As a result, the preparation of the appointment of the employee as a data processor is facilitated.

This method allows us to easily prepare records of processing activities and to increase the accountability of an organization to the GDPR standards.

The ability to demonstrate that the records of processing activities and the list of data processors derived from the modelling and analysis of business processes can certainly increase the organization's accountability and credibility.

We are developing PDARC, the add-on of Eclipse BPMN2 Modeler, an Eclipse Plugin that implements the most important BPMN feature to add GDPR information to the BPMN model. PDARC is also able to automatically extract the records of processing activities and the roles of the data processor by querying the BPMN modelling system.

We also aim to apply this approach in different contexts. We are working with institutions in the Local Public Administration, and in the Health Care context to help manage special categories of personal data that are the object of a specific prescription, Article 9 of GDPR.

In the future, we aim at creating a tool that can be integrated with a company's run time workflow engine to automatically generate records of processing activities and the appointment of data processors that are adjusted with the requirements of company processes.

Acknowledgements This work was partially supported by the "EASYPAL" project no. Y95B457, funded by Innolabs Apulia Region (Italy), entitled "Sostegno alla creazione di soluzioni innovative finalizzate a specifici problemi di rilevanza sociale" under the POR Puglia FESR 2014-2020 research program sub-azione 1.4.B.

References

1. E.-O. Wilhelm, A brief history of the General Data Protection Regulation
2. M. Pesic, H. Schonenberg, W.M.P. Van Der Aalst, DECLARE: full support for loosely-structured processes, in *Proceedings—IEEE International Enterprise Distributed Object Computing Workshop, EDOC* (2007)
3. T.T. Hildebrandt, R.R. Mukkamala, Declarative event-based workflow as distributed dynamic condition response graphs (2011). arXiv:1110.4161
4. D. Harel, M. Politi, I. Books24x7, *Modeling Reactive Systems with Statecharts* (1998)
5. A.M. Fernandez-Saez, D. Caivano, M. Genero, M.R.V. Chaudron, On the use of UML documentation in software maintenance: results from a survey in industry, in *2015 ACM/IEEE 18th International Conference on Model Driven Engineering Languages and Systems, MODELS 2015—Proceedings* (2015), pp. 292–301
6. O.M. Group, OMG unified modeling language TM (OMG UML), Superstructure v.2.5. *InformatikSpektrum* (2015)
7. P. Ardimento, D. Caivano, M. Cimitile, G. Visaggio, Empirical investigation of the efficacy and efficiency of tools for transferring software engineering knowledge. J. Inf. Knowl. Manag. **7**(3), 197–207 (2008)
8. S. España, N. Condori-Fernandez, A. González, O. Pastor, An empirical comparative evaluation of requirements engineering methods. J. Braz. Comput. Soc. **16**(1), 3–19 (2010)
9. R. Hull et al., Introducing the guard-stage-milestone approach for specifying business entity lifecycles, in *Lecture Notes in Computer Science (including subseries Lecture Notes in Artificial Intelligence and Lecture Notes in Bioinformatics)* (2011)

10. Object Management Group, *Case Management Model and Notation (CMMN)* (2013)
11. Object Management Group, *Business Process Model and Notation (BPMN) Version 2.0* (2011)
12. M. Cremonini, E. Damiani, S.C. di Vimercati, P. Samarati, A. Corallo, G. Elia, *Security, Privacy, and Trust in Mobile Systems and Applications* (IGI Global, 2005)
13. M. Enamul Kabir, H. Wang, E. Bertino, A conditional purpose-based access control model with dynamic roles. Expert Syst. Appl. (2011)
14. C. Ardito, U. Barchetti, A. Capodieci, A. Guido, L. Mainetti, Business process design meets business practices through enterprise patterns. Int. J. e-Collab. **10**(1), 57–73 (2014)
15. U. Barchetti, A. Capodieci, A.L. Guido, L. Mainetti, Modelling collaboration processes through design patterns. Comput. Inf. **30**(1), 113–135 (2011)
16. A. Capodieci, L. Mainetti, L. Alem, An innovative approach to digital engineering services delivery: an application in maintenance, in *2015 11th International Conference on Innovations in Information Technology (IIT) (IIT'15)*, Dubai, UAE (2015), pp. 336–343
17. W. Labda, N. Mehandjiev, P. Sampaio, Modeling of privacy-aware business processes in BPMN to protect personal data, in *Proceedings of the 29th Annual ACM Symposium on Applied Computing*, New York, NY, USA (2014), pp. 1399–1405
18. A.D. Brucker, Integrating security aspects into business process models. Inf. Technol. **55**(6), 239–246 (2013)
19. Y. Cherdantseva, J. Hilton, O. Rana, Towards SecureBPMN—aligning BPMN with the information assurance and security domain, in *Business Process Model and Notation* (2012), pp. 107–115
20. O. Altuhhov, R. Matulevičius, N. Ahmed, An extension of business process model and notation for security risk management. Int. J. Inf. Syst. Model. Des. (IJISMD) **4**(4), 93–113 (2013)
21. P. Pullonen, R. Matulevičius, D. Bogdanov, PE-BPMN: privacy-enhanced business process model and notation, in *Business Process Management* (2017), pp. 40–56
22. M. Salnitri, F. Dalpiaz, P. Giorgini, Designing secure business processes with SecBPMN. Softw. Syst. Model. **16**(3), 737–757 (2017)
23. M.E.A. Chergui, S.M. Benslimane, A valid BPMN extension for supporting security requirements based on cyber security ontology, in *Model and Data Engineering* (2018), pp. 219–232
24. C.L. Maines, D. Llewellyn-Jones, S. Tang, B. Zhou, A cyber security ontology for BPMN-security extensions, in *2015 IEEE International Conference on Computer and Information Technology; Ubiquitous Computing and Communications; Dependable, Autonomic and Secure Computing; Pervasive Intelligence and Computing* (2015), pp. 1756–1763
25. C.L. Maines, B. Zhou, S. Tang, Q. Shi, Adding a third dimension to BPMN as a means of representing cyber security requirements, in *2016 9th International Conference on Developments in eSystems Engineering (DeSE)* (2016), pp. 105–110
26. K.S. Sang, B. Zhou, BPMN security extensions for healthcare process, in *2015 IEEE International Conference on Computer and Information Technology; Ubiquitous Computing and Communications; Dependable, Autonomic and Secure Computing; Pervasive Intelligence and Computing* (2015), pp. 2340–2345
27. A. Rodríguez, E. Fernández-Medina, M. Piattini, A BPMN extension for the modeling of security requirements in business processes. IEICE Trans. Inf. Syst. (2007)
28. M. Menzel, I. Thomas, C. Meinel, Security requirements specification in service-oriented business process management, in *2009 International Conference on Availability, Reliability and Security* (2009), pp. 41–48
29. M. Palmirani, M. Martoni, A. Rossi, C. Bartolini, L. Robaldo, PrOnto: privacy ontology for legal reasoning, in *Electronic Government and the Information Systems Perspective* (2018), pp. 139–152
30. D. Basin, S. Debois, T. Hildebrandt, On purpose and by necessity: compliance under the GDPR, in *Financial Cryptography and Data Security (FC)* (2018)
31. A. Capodieci, L. Mainetti, Business process awareness to support GDPR compliance, in *In Proceedings of the 9th International Conference on Information Systems and Technologies (ICIST 2019)*. ACM, New York, NY, USA, Article 2 (2019), 6 pages. https://doi.org/10.1145/3361570.3361573

Chapter 17
Mapping Design Anthropology: Tracking the Development of an Emerging Transdisciplinary Field

Christine Miller and Ken Riopelle

Abstract The practice of design anthropology has continued to evolve since the publication of *Design + Anthropology: Converging Pathways in Anthropology and Design* in 2018. At that time, design anthropology was described as "an emerging transdisciplinary field." ([1], [2]: 10, [3]). Working collaboratively with Ken Riopelle who provided analytical expertise in social network analysis, we approached this claim from the perspective of social network analysis "to investigate the human and nonhuman actors (i.e., people and institutions) that have contributed to design anthropological practice and theorizing." [3]. Our initial goal was to determine if—and, if so, to what extent—design anthropology qualified as a disciplinary "field". In our original analysis, we began by establishing a set of benchmarks that serve as indicators to identify a disciplinary field. In this paper, we revisit our initial analysis, updating it with new publications, contributors, blogs, groups, and other developments, to investigate if and how design anthropology has diffused.

Introduction

Design anthropology emerged in the late 1970s as developments within business and industry, labor, and the government created opportunities for collaboration between designers and anthropologists ([4]: 5). Design anthropology has been described alternately as "an emerging transdisciplinary field", "a (sub)discipline", and as "a fast-developing academic field that combines elements from design and anthropology." However, these claims had not been substantiated leaving questions as to the actual size, presence, and potential impact of design anthropology within the broad category of established disciplinary fields. Was design anthropology a practice, a research

C. Miller (✉)
Savannah College of Art and Design, Savannah, USA
e-mail: millerphd08@gmail.com

K. Riopelle
Wayne State University, Detroit, USA
e-mail: kenriopelle@me.com

© Springer Nature Switzerland AG 2020
A. Przegalinska et al. (eds.), *Digital Transformation of Collaboration*,
Springer Proceedings in Complexity,
https://doi.org/10.1007/978-3-030-48993-9_17

strategy, a subject area, or a discipline? Or was it a subfield of business anthropology, as it is often categorized in the U.S.? The fact that these questions can only be answered from a specific perspective that creates a snapshot of its presence in a particular time and place does not diminish their importance. Mapping and tracking the *People* and *Events* that are directly influencing the diffusion of design anthropology practice and theory enables us to see whether (or not) design anthropology is gaining traction in the marketplace of ideas. Visualizing networks of people and things (e.g., programs, institutions, conferences) directly related to design anthropology allows us to monitor if and how it is becoming a recognizable field in its own right.

Working collaboratively with Ken Riopelle who provided analytical expertise in social network analysis, we approached this claim from the perspective of social network analysis "to investigate the human and nonhuman actors (i.e., people and institutions) that have contributed to design anthropological practice and theorizing." [3].

What Constitutes a Disciplinary Field?

We began by establishing a basis for design anthropology as a unique branch of knowledge production, creating a set of benchmarks to serve as indicators in answering the question "What constitutes a disciplinary field?" In this paper we revisit our initial analysis, updating it with new publications, contributors, blogs, groups, and other developments, to investigate if and how design anthropology has diffused. To establish a basis for design anthropology as an "emerging disciplinary field", we turned to recognized sources for guidance [3]:

> A more rigorous formal classification method that is used to determine if and when a field of study qualifies as a discipline is based on an extensive search of "citable items". For example, Thompson and Reuters' Web of Science (WoS) is a widely recognized resource that provides this type of formal analysis.[1] The Web of Science (WoS) tracks the emergence of new subject categories (SC) based on "citable items" that include journal articles, conference proceedings, and reviews [5]. The Science and Social Science Index (SCI + SoSCI), which is updated periodically, currently includes 252 subject categories (SC) across six broad areas (Table 4-1). Classifying and cataloging emerging subject areas is a complex ontological task. Not only is the terminology confusing ([5]: 589–590), but also because the process is dynamic. The rate of new subject areas is steadily increasing as science becomes more interdisciplinary [6]. Some subject categories, for example, Chemistry, are split off into separate categories while others may be eliminated.

Because design anthropology was not among the lists of recognized fields, we conceptualized the process of becoming a recognized field as progressing along a

---

[1]Thompson Reuters' *Web of Science* (Accessed July 12, 2016) http://ipscience.thomsonreuters.com/.

trajectory.[2] In this case, we needed to operationalize the concept of *discipline* by identifying a set of reliable indicators, which we did in the 2018 publication.[3]

Operationalizing Design Anthropology

Our task was to operationalize the concept of "discipline", which we accomplished by establishing a set of indicators that identified "dedicated conferences and seminars, funding and sponsorship, journals, research agendas, recognized experts, professional societies and organizations, academic courses and programs focused on the subject area, and dissertations that specifically focus on the subject area." [3].

We began to compile lists for each indicator by using several readily available web-based analytic tools to collect data from the Internet that would provide a broad-brush preliminary view of the domain design anthropology.[4] We used Google Books Ngram Viewer, which displays a graph showing how a phrase has occurred in a corpus of books over a selected time range between 1880 and 2008.[5] We searched on both "design anthropology" and "design and anthropology". We knew of no books on design anthropology published prior to 2008. Ngram confirmed this when the search on "design anthropology" yielded no results. However, entering "Design" and "Anthropology" as separate terms yielded the results shown in Fig. 17.1. Note that "Design" occurs more often than "Anthropology", especially since 1960 [3].

By way of comparison, a search on the terms "social network analysis" and "network analysis" within the same time period yielded results depicted in Fig. 17.2.

To broaden our search, we next conducted a series of searches using Google Scholar using multiple terms including "design anthropology", "design + anthropology", and "design and anthropology". Again, we recognized that Ngram and Google Scholar searches will produce different results based on factors such as the date on which they're run.[6] However, they can provide an estimate of the occurrence of the term in articles and other publications.

Finally, we ran an Ngram search comparing "Anthropology", "Business Anthropology", and "Design Anthropology" (Fig. 17.3) that shows the topic "Anthropology" peaking before 2000 and "Business Anthropology" barely up from zero. Design Anthropology does not show up as a topic at all. Taken together these graphs beg the question as to *why* Business Anthropology and Design Anthropology have such a

---

[2]We do not envision this trajectory as a linear progression, rather we see it as nonlinear with "progress" marked by ebbs and flows of events and periods of activity.

[3]Refer to Chap. 4 in *Design + Anthropology: Converging Pathways in Anthropology and Design* [3].

[4]For a detailed explanation of the Ngram searches please refer to pp. 78–80 in *Design + Anthropology: Converging Pathways in Anthropology and Design* [3].

[5]Google Ngram (Accessed July 12, 2016) https://books.google.com/ngrams.

[6]Data is continuously being added. These searches on Google Scholar were conducted on July 12, 2016.

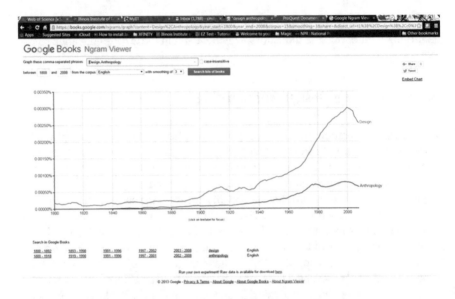

Fig. 17.1 Ngram results for "Design" and "Anthropology" (1800–2008)

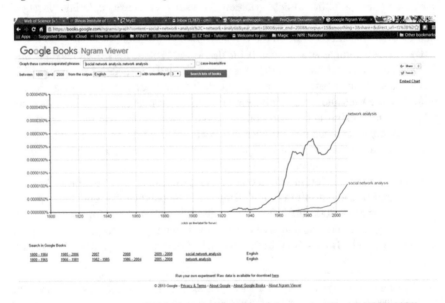

Fig. 17.2 Ngram comparing "network analysis" and "social network analysis" (1800–2008)

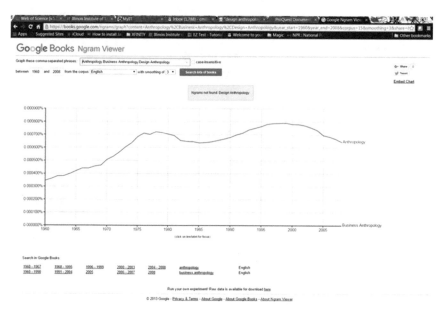

Fig. 17.3 Ngram comparing "Anthropology", "Business Anthropology", and "Design Anthropology" (1960–2008)

weak showing. As relatively new areas of scholarship and practice this is understandable. But for Anthropology as a long-established discipline the decline in publications after 2000 is somewhat more puzzling. We might hypothesize that much of the literature that documents the work of anthropologists is being subsumed into the literature of different disciplines, but this is purely speculation.

To complete our broad-brush Internet query, we conducted searches on ProQuest[7] using the same terms. Mapping the results of these high-level searches to the indicators we had established to operationalize the concept of "field" suggested that "as a unique field of knowledge production design anthropology has not achieved the level of subject category or discipline." [3]. However, there was evidence to suggest that design anthropology was an emerging *area of interest* that had the potential to grow. Our next step was to visualize the networks that were forming around the practice of design anthropology.

---

[7]For a detailed explanation of the Google Scholar and ProQuest searches in the original research refer to pp. 76–82 in *Design + Anthropology: Converging Pathways in Anthropology and Design* [3].

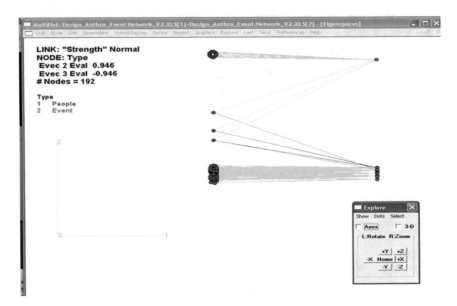

Fig. 17.4 MultiNet projection of the 180 people contributing to 12 design anthropology events from 2011 to 2016

Mapping Design Anthropology

By 2016, there were multiple books, articles, conferences, presentations, and academic programs that were directly related to design anthropology. We began organizing data by creating a spreadsheet[8] that included a list of 12 discrete "Events"—edited books, seminars, committees, networks, conferences, and conference panels—explicitly related to design anthropology between 2011 and 2016. We added a worksheet for "People" and listed the names of each individual who contributed to or participated in each of these 12 events arriving at a total of 180 individuals (i.e., discrete nodes). This list allowed us to create network views by exporting the names to several dynamic network analysis tools including MultiNet and NEGOPY. Figures 17.4 and 17.5 depict the results. Figure 17.4 shows the distribution of 150 People (blue) over 12 Events (red) that were identified as "design anthropological" explicitly by their title or name of publication. Especially interesting in Fig. 17.5 was the emergence of three individuals who emerged as "liaisons" between the 2011 edited collection **Design Anthropology: Object Culture for the 21st Century**, the core group of 11 remaining Events, and the 52 People who participated in two or more events. We interpreted these 52 individuals to comprise a significant core.

---

[8]The spreadsheets, including a complete list of URLs and the search results, were made available on the companion website for *Design + Anthropology: Converging Pathways in Anthropology and Design.*

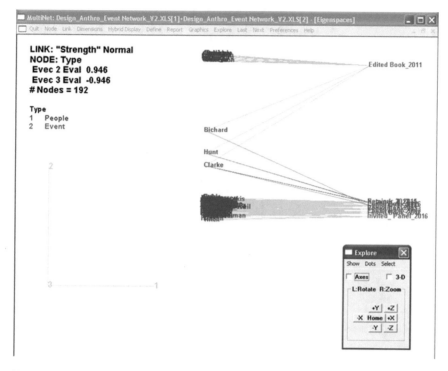

Fig. 17.5 MultiNet projection of the 180 people contributing to 12 design anthropology events from 2011 to 2016, with names

Updating Design Anthropology Networks

With the 2016 data as a benchmark, we are now interested in investigating how design anthropology has fared over the 3 years since the initial study. Has design anthropology grown or stagnated? Have there been additional books, dedicated conferences and seminars, funding and sponsorship, journals, research agendas, recognized experts, professional societies and organizations, academic courses and programs focused on the subject area, and dissertations that specifically focus on design anthropology? Have new networks emerged? In other words, by revisiting the indicators, can we conclude that design anthropology is moving closer to becoming a defined field or subfield?

Table 17.1 shows 8 new Events plus 27 publications that have occurred since 2016 and two Events that were initiated in 2014–2015. Two LinkedIn groups and a SLACK network were not included in the 2016 research. These are large groups that are composed of people with varying degrees of interest in and knowledge of design anthropology.

Table 17.1 New events since original research was conducted

Year/Event	# People
2015—LinkedIn Design Anthropology group: Brandon Meyer, Owner	338
2014—LinkedIn DesAnthro group: Natalie Hanson, owner (companion to Anthrodesign group)	464
2018—Designs and Anthropologies Colloquium (SAR seminar)	9
2018—*Journal of Business Anthropology* (Special Issue on Design Anthropology)	14
2018—**Anthropology Designed** (working title; available late 2019): Contributors	7
2018—"Design by Anthropologists" (NAPA blog)	4
2018—**Design Anthropology: Object Cultures in Transition**	16
2019—"Design Anthropology" (Oxford Research Encyclopedia Anthropology)	1

Discussion

The purpose of this research is to follow up on the earlier work conducted by Christine Miller and Ken Riopelle in the chapter "Mapping Design Anthropology" from the book **Design + Anthropology: Converging Pathways in Anthropology and Design** [3]. The research conducted in 2016 and published in the 2018 text used 12 selected Events to map participation in Design Anthropology with a headcount of 295 people [3]. The current work (i.e., 2019) added an additional 16 Events as for a total of 28 events and a new headcount of 1,183 people. Table 17.2 shows a combined list of the original Events and new Events. Network analysis using the MultiNet/NEGOPY program produced three distinct groups that are color-coded below.

The Standard Distance is a network statistic that indicates the degree of centrality for a node (person or list). *The greater the negativity score the stronger a node's connection is within the group.*

The Events are sorted by color and by their standard distance score.

Figure 17.6 is a simple mapping of the 28 Events (i.e., lists) and the corresponding names of individuals. The Events are colored in red on the left and the names are colored in blue on the right. *The vertical distance between the nodes (lists and name) is meaningful and represents how far apart the nodes are from each other.* The nodes at the top are far away or not connected to those nodes at the bottom. A tight cluster of nodes means they are close or connected to one another because of overlapping list memberships.

Figure 17.7 shows Design Anthropology Groups color the nodes by groups found with the MultiNet/NEGOPY programs. The Map Key indicates 4 distinct groups with the colors: blue, red, green, and purple. Note that the Group 0 (Blue) is technically considered a tree node and not a separate defined group. This tree node consists of 4

Table 17.2 Three groups of combined events produced by network analysis

List #	List Name	Alias	Count	Negopy Group	Color	Standard Distance
1 L11 - Design Anthropology: Object Culture for the 21st Century (2011)		15	19	1	Red	-1.764
24 L18b - Design Anthropology: Object Cultures in Transition: Contributors		24	16	1	Red	-1.764
17 L13 - LinkedIn Design Anthropology group: Brandon Meyer, Owner		17	335	2	Green	-2.399
18 L14 - Linked in DesAnthro group		18	460	2	Green	-2.399
26 L19b - Design by Anthropologists: Contributors		26	2	2	Green	2.435
10 L7A - Design Anthropological Futures Conference Copenhagen, DK Aug. 13-14, 2015: Position Papers		8	58	3	Purple	-1.851
4 L2 - Research Network for Design Anthropology: Participants		2	30	3	Purple	-1.626
13 L8B - Design Anthropological Futures (available Nov. 2016): Contributors		11	28	3	Purple	-1.288
7 L4 - Seminar 2: Interventionist Speculation (August 2014): Presenters		4	25	3	Purple	-1.120
9 L6B - Design Anthropology: Theory and Practice (2013): Contributors		7	17	3	Purple	-0.951
8 L5 - Seminar 3: Collaborative Formation of Issues (January 2015): Presenters		5	19	3	Purple	-0.389
6 L3 - Seminar 1: Ethnographies of the Possible (April 2014): Presenters		3	16	3	Purple	-0.276
28 L21 Google Scholar Search on September 3, 2019 N=22		28	35	3	Purple	-0.220
19 L15 - Designs and Anthropologies Colloquim		19	9	3	Purple	-0.051
22 L17b - Anthropology Designed (working title; available late 2019): Contributors		22	7	3	Purple	0.005
2 L12 - Design and Anthropology (2012)		16	21	3	Purple	0.061
14 L8C - Design Anthropological Futures (available Nov. 2016): Reviewers		12	4	3	Purple	0.061
15 L9 - Anthropology + Design Mutual Provocations Conference. San Diego, Oct. 27-29, 2016: Invited Speakers		13	13	3	Purple	0.061
3 L1 - Research Network for Design Anthropology: Steering Committee		1	7	3	Purple	0.118
12 L8A - Design Anthropological Futures (available Nov. 2016): Editors		10	6	3	Purple	0.174
20 L16- Journal of Business Anthropology (Special Issue on Design Anthropology)		20	9	3	Purple	0.343
5 L6A - Design Anthropology: Theory and Practice (2013): Editors		6	3	3	Purple	0.455
16 L10 - AAA Annual Meeting Minneapolis, Nov. 16-20, 2016 Design Anthropology Panel		14	8	3	Purple	0.680
11 L7B - Design Anthropological Futures Conference Copenhagen, DK Aug. 13-14, 2015: Posters		9	31	3	Purple	0.736
21 L17a - Anthropology Designed (working title; available late 2019): Editors		21	2	3	Purple	1.242
25 L19a - Design by Anthropologists: Terry Redding, Owner		25	1	Dyad		
23 L18a - Design Anthropology: Object Cultures in Transition: Editor		23	1	Isolate		
27 L20 - Design Anthropology (Oxford Research Encyclopedia Anthropology - OREA)		27	1	Isolate		
			1183			

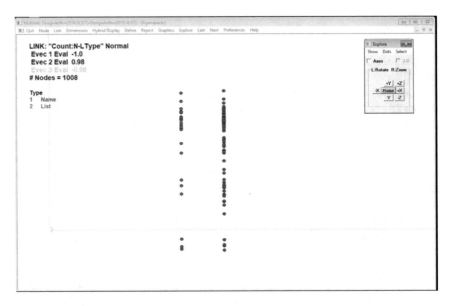

Fig. 17.6 Lists and names

nodes from List #14 with one connection to the larger Group 3. Thus, it appears at the very top of the vertical rankings and is distant from all the other nodes.

Figure 17.8 shows the Design Anthropology Groups with the individual's names. The event lists are ordered vertically on the left and the names are on the right. Again, we can see the four clear groups by their color. The next series of maps will provide a brief description of each group.

Figure 17.9 shows Group 1 (Red) and Group 2 (Green) with members identified in the panel at the left. There are 9 members in Group 1 (Red). List 1 and List #24 are associated with this group. Group 2 (Green) has 19 members and includes the two LinkedIn lists; node List #18 by Natalie Hanson and node List #17 by Brandon Meyer. These two groups are very far apart in the network mapping with the Red Group near the top and Green group at the very bottom. *There are no connections between them.* Note there are few connections between these two large online LinkedIn lists. There are connections from Group 2 (Green) to Group 3 (Purple) but there are no connections near the top of Group 3, with the exception of Christine Miller, who is connected to List #10. It appears these are disconnected Design Anthropology memberships. For a complete listing of L17 and L18, see attached.

Figure 17.10 shows that Group 2 (Purple) has 77 members. The listing of the Group 2 extends over two frames. The strength of the members' connections is reflected in their standard distance score. *The more negative score the stronger their connection.*

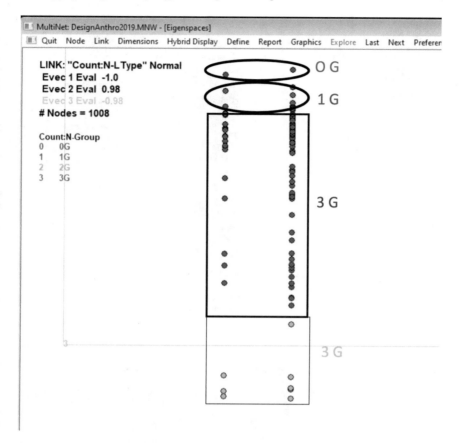

Fig. 17.7 Design anthropology groups

Conclusion and New Developments

For this update, we invited Wendy Gunn, Jinshan Distinguished Professor at Jiangsu University and author and lecturer on design anthropology, to review our paper. Her comments and questions were insightful, especially in highlighting the need to balance Automated Social Network Analysis (ASNA) and web-based research with the grounded ethnographic inquiry. For example, in response to Gunn's question as to whether we referred to classic texts concerning the development of disciplinary fields in terms of Western knowledge production, we expanded our research to include the trend toward more interdisciplinary science that is supported in the disciplinary development and composition literature [5–7]. Describing techniques to apply science overlay maps, Rafols et al. illustrate how these maps "help investigate the increasing number of scientific developments and organisations that do not fit within traditional disciplinary categories."

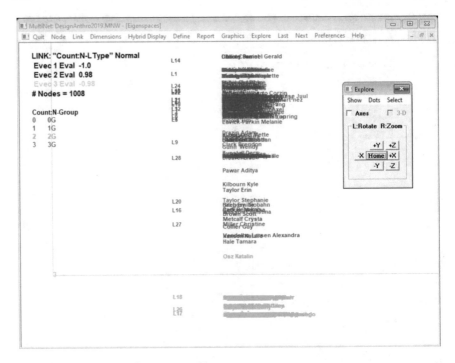

Fig. 17.8 Design anthropology groups with names

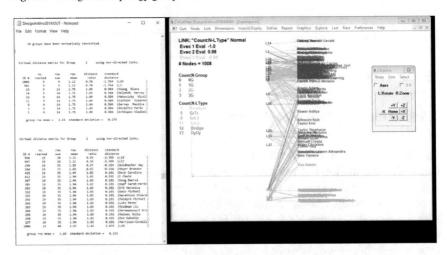

Fig. 17.9 Group 1 (Red) and Group 3 (Green) with members

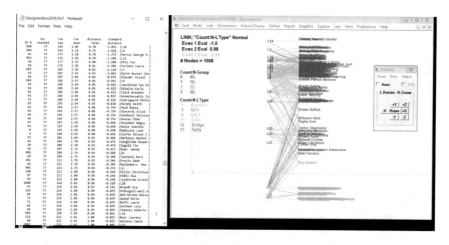

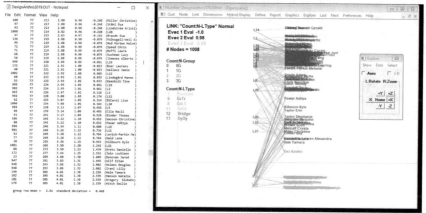

Fig. 17.10 Group 2 (Purple)

Most science and technology institutions have undergone or are undergoing major reforms in their organisation and in their activities in order to respond to changing intellectual environments and increasing societal demands for relevance. As a result, the traditional structures and practices of science, built around disciplines, are being by-passed by organisation in various ways in order to pursue new types of differentiation that react to diverse pressures (such as service to industry needs, translation to policy goals, openness to public scrutiny, etcetera). However, no clear alternative socio-cognitive structure has yet replaced the "old" disciplinary classification. In this fluid context, in which social structure often no longer matches with the dominant cognitive classification in terms of disciplines, it has become increasingly necessary for institutions to understand and make strategic choices about their positions and directions in moving cognitive spaces. "The ship has to be reconstructed while a storm is raging at sea." (Neurath, 1932/33).

In our initial question concerning the status of design anthropology as a discipline, we did not consider the trend toward interdisciplinary science. We were only asking *if* design anthropology could be classified as a disciplinary field. While the answer based on formal orders of classification is "not at this time", our perspective has shifted: we now see the evolution of design anthropology as part of a higher level trend toward interdisciplinary science. Referencing her lecture at Carnegie Mellon's School of Design in 2019, Gunn alluded to this trend in her comments, suggesting that "Design Anthropology among other emerging fields is one area of knowledge production which is contributing to a yet unnamed discipline across different fields and sectors."[9]

Emerging trends in the "design-and-anthropology" relationship support this observation and suggest that design anthropology might be overcome by new developments within anthropology and especially within the field of design. Although it appears that design anthropology is an established practice in European, especially in Scandinavian countries, this does not appear to be the case in the U.S. Rather than design anthropology growing as a distinct field of knowledge production, we wonder if the focus has turned *inward* instead of *across* disciplines. With the exception of the anthropology department at the University of North Texas with interdisciplinary design and "project-based anthropology", it appears that the diffusion of design anthropology as a distinct program has stalled on the institutional level. However, at least two U.S. institutions recently granted doctoral degrees with "Design Anthropology" in the individual's title.[10]

While the trend toward interdisciplinary science is increasing, there also seems to be a counter-trend toward disciplinary retrenchment. For example, within Design in the U.S., we might be seeing a retrenchment in which Design is incorporating aspects of social science (i.e., anthropological) theory and methodology into its core. This is reflected in new subfields like "transitional design", "design research", and "social design".

In terms of conferences, Ethnographic Praxis in Industry (EPIC) continues to be a strong draw for designers, but perhaps less so for anthropologists who might identify more with the subfield of business anthropology.[11] It is interesting to note that the American Anthropological Association (AAA) annual meetings and the Society for Applied Anthropology (SfAA) have many workshops and panels focused on "design". However, at design conferences in the U.S., it is less likely to hear anthropology mentioned.[12] With a dearth of academic opportunities in the U.S., young anthropologists and anthropologists seeking to transition from academic careers are continuing to find their way into practice within the private and public sector, especially in the area of User Experience (UX).

---

[9]Personal communication with Wendy Gunn in December 2019.

[10]A cursory Google search (December 18, 2019) revealed ten individuals with a Ph.D. in Design Anthropology. Two of the degrees were from U.S. institutions (Duke University and the University of Texas, Austin); most were from institutions outside the U.S. including Curtain University in Australia, Sheffield Hallam University in the UK, Mads Plank Institute at the University of Southern Denmark.

[11]Refer to the Business Anthropology website https://www.businessanthro.com/ and Global Business Anthropology Summit https://www.businessanthro.com/2019-summit

[12]It was pointed out by a reviewer that this is not true in the European context, who noted that anthropologists often participate in design conferences.

We would like to continue this analysis by exploring questions such as: Why are the LinkedIn Design Anthro and Design Anthropology Groups nonoverlapping? *Why aren't these groups connected?* Do they have different orientations? What would explain this divergence? Or is this mapping just an artifact and the members are connected and collaborate more that appears from these maps? Is it possible or desirable to attempt to create more overlap between these groups? Finally, given the trend toward interdisciplinary science, we will expand our inquiry to consider the evolution of Design Anthropology, as Gunn suggests, "as one area of knowledge production among other emerging fields which is contributing to a yet unnamed discipline across different fields and sectors.".

References

1. P. Ehn, E.M. Nilsson, R. Topgaard (eds.), *Making Futures: Maringal Notes on Innovation, Design and Democracy* (MIT Press, Cambridge, MA, 2014)
2. W. Gunn, T. Otto, R.C. Smith (eds.), *Design Anthropology: Theory and Practice* (Bloomsbury, New York, 2013)
3. C. Miller, *Design + Anthropology: Converging Pathways in Anthropology and Design* (Taylor & Francis, New York, 2018)
4. T. Otto, R.C. Smith, A distinct style of knowing, in *Design Anthropology*, eds by W. Gunn, T. Otto, R.C. Smith. Theor. Pract. (Bloomsbury, New York, 2013) 1–29
5. L. Leydesdorff, S. Carley, I. Rofols, Global maps of science based on the new web of science categories. Sociometrics **94**, 589–593 (2013)
6. A. Porter, I. Rafols, Is science becoming more interdisciplinary? Measuring and mapping six research fields over time. Sociometrics **81**(3), 719–745 (2009)
7. I. Rafols, A.L. Porter, L. Leydesdorff, Science overlay maps: a new tool for research policy and library management. J. Am. Soc. Inform. Sci. Technol. **61**(9), 1871–1887 (2010)

Chapter 18
Combining Social Capital and Geospatial Analysis to Measure the Boston's Opioid Epidemic

Cordula Robinson, Michael Wood, Francesca Grippa, and Earlene Avalon

Abstract Social support is considered an important factor in the recovery of individuals, who suffer from drug use disorder. Traditional drug treatment interventions have mainly focused on the individual without taking into consideration the social and environmental conditions that may support or reduce drug use. By combining a social capital framework with geospatial research methodologies, we mapped hot spots and cold spots within the 23 Boston neighborhoods and identified where social ties were stronger or weaker. The spatial correlation analysis and Geographically Weighted Regression demonstrated that in areas where social capital is low, there is a moderately high incidence of opioid deaths and sick assist calls. Our analysis shows that in neighborhoods where residents are involved in charitable organizations, where people gather around religious organizations, or where unions are more active, people help each other more and might be aware of actions to take to prevent opioid-related deaths.

Introduction

Opioid-related deaths have increased dramatically over the past few years, and the opioid-related death rate in Massachusetts is now more than twice the national rate. According to the 2017 report of the Government of Massachusetts Department of Public Health, about one-third of admissions to substance abuse treatment centers and programs in Massachusetts are opioid-related (2017). The opioid epidemic has been exacerbated in recent years due to decades of opioid over-prescription, and the influx of cheap heroin, and the emergence of fentanyl [1]. Recent studies have used opioid overdose and fentanyl-related fatality data between 2015 and 2018 to identify neighborhoods in Boston with high densities of opioid incidents. Robinson and colleagues inspected the relationship between opioid reversal drugs and fentanyl

C. Robinson · M. Wood · F. Grippa (✉) · E. Avalon
Northeastern University, 360 Huntington Avenue, Boston, MA, USA
e-mail: f.grippa@northeastern.edu

© Springer Nature Switzerland AG 2020
A. Przegalinska et al. (eds.), *Digital Transformation of Collaboration*,
Springer Proceedings in Complexity,
https://doi.org/10.1007/978-3-030-48993-9_18

deaths and found that lifesaving opioid reversal drugs are less accessible to some residents than others [2].

In this study, we combined social capital variables and geospatial analysis to identify social factors that contribute to building resilience within a community where hot spots are identified. Data-driven approaches help to focus the attention on historically underrepresented and overlooked areas. In this study, we combine geospatial analysis and social capital analysis to better understand the Boston's opioid problem. Our study builds on previous work that applied spatial analyses using Geographic Information Systems (GIS) to identify significant clusters of fatal overdoses through discarded syringes [3, 4].

Most of the traditional drug treatments and political decisions ordinarily consider the individual without taking into account social and environmental conditions. The assumption of our study is that we can measure the resilience of communities to the opioid crisis by observing the social connections within those communities, which can lower the social stigma and provide empowerment opportunities for the individual. We believe that the social fabric of a community may enhance a community's ability to respond to the crisis, by complementing drug treatment options and social support within treatment facilities [5].

Community resilience has been associated with various adaptive capacities, including economic development, social capital, information and communication infrastructure, and community competence [6]. In this study, we correlate social capital metrics with opioid incident data (sick assists and death) for the 23 Boston neighborhoods and apply the social capital framework defined by Aldrich [7]. We integrate variables for bonding, bridging, and linking social capital at a US Census Block group level to examine community opioid resiliency through the lens of social ties. The goal is to appraise a community's resilience to the opioid crisis by its ability to adapt to this social emergency in the context of its intrinsic social makeup.

Social capital is usually conceptualized as being embodied in the social ties among individuals and their positions and can be measured by resources available to individuals via interpersonal ties and institutional connections, including family, school, and work [8, 9]. A community can support individuals by engaging in various types of support, from instrumental—e.g., providing shelter, food, access to job opportunities—to emotional—e.g., creating or sponsoring charitable organizations or supporting religious affiliations.

In the context of the opioid crisis, embeddedness in a supportive pro-social network—represented by family members, non-drug-using friends, and friends who have recovered from drug addiction—might be expected to mitigate the crisis, including facilitating recovery and making individuals feel accepted in their community. Conversely, being associated with a drug user network can generate negative social capital and reinforce the social identity of marginalization and strong public discrimination, exacerbating conditions [10].

Previous studies show that the possession of positive social capital greatly enhances the likelihood to reduce the risk level of drug use, whereas the possession of negative social capital reduces such likelihood [11]. In a study on social network characteristics associated with risky behaviors among runaway and homeless youth, Ennet and colleagues found that youth without a social network were significantly more likely to report current illicit drug use [12]. Their results indicate that networks had risk-enhancing and risk-decreasing properties in that network characteristics were associated in both positive and negative directions with risky behaviors. For example, in a study focused on the impact of adverse family circumstances on subsequent deviance, Nurco and colleagues [13] found that disruption in the family structure was significantly associated with crime severity.

A qualitative study conducted among homeless individuals with substance use disorders found that family members, in particular mothers, play a key role in the decision to seek treatment, thanks to dynamics such as confrontation and ongoing emotional support [14]. Members of the recovery network provided empathic emotional support; coworkers, outside friends, health professionals, and romantic relationships were also mentioned as important factors determining a successful recovery.

The literature on "negative social networks" tells us that having close friends who are still using drugs could reduce the ability to recover. Latkin and colleagues [15] found that several social network variables, including a larger number of conflictual ties among the network members and a larger number of network members who were injection drug users, were significantly associated with drug overdose in the prior 2 years. Another study on peer, family, and motivational influences on drug treatment process and recidivism found that peer deviance was positively related to re-arrest [16].

As demonstrated by two research studies in the Chicago area and in Stockholm, Sweden [17], the economic disadvantage in urban areas and stability are good predictors of collective efficacy. In areas where there are more of these social gathering opportunities (including narcotics anonymous and alcoholics anonymous), individuals engage in social support activities and look out for others, and help individuals struggling with health problems such as opioid addiction or mental illness [18]. In geographic areas where there is a high degree of bonding and bridging social capital, individuals are involved in behaviors of "reciprocal exchange", doing favors for each other, practicing acts of kindness, and lending a hand whenever possible [18].

Social epidemiology studies focused on individuals with substance use disorders [19] have demonstrated the importance of a relational approach, where the social context has a strong influence on substance use patterns. For example, with reference to smoking behavior, in a 26-year study of African Americans in inner-city Chicago, Illinois, those with poorer relationships with their family were more likely to start smoking. Smoking behavior of social network members has also been shown to be important determinants of age at smoking initiation. In a prospective cohort study of 996 adolescents in Sydney, Australia, followed for 1 year, characteristics of one's social networks (particularly drug use in the social network) were associated with the likelihood of initiating marijuana use.

Research Design

Operational Framework and Hypotheses

Aldrich's classification of social capital in its bonding, bridging, and linking components, is a helpful framework to identify social resiliency. We mapped social ties in the hot spots and cold spots areas by differentiating between strong connections between individuals struggling with drug use disorder and others who are emotionally close to them (bonding), as well as potential acquaintances or individuals loosely connected (bridging) and ties to individuals in power (linking) [8, 20]. Bonding is often described as "*good will, fellowship, mutual sympathy, and social intercourse among a group of individuals and families who make up a social unit*" [21]. Sociologists call this phenomenon homophily to indicate how closest friends and contacts likely share language, ethnicity, culture, and class [22].

Bridging refers to the type of connection that comes from weaker ties to people with whom we spend less time and have less in common. The connections may come through a church and other community meeting points [23]. As demonstrated by the seminal work by Granovetter [24], less intimate connections between people based on more infrequent social interaction, which Granovetter calls "weak ties", are critical to access social resources, such as job referrals, because they integrate the community by bringing together otherwise disconnected subgroups [18]. In some communities strong ties among neighbors are no longer the norm, as friends and social support networks are decreasingly organized around local institutions, either religious or civic. Bridging ties may be especially useful during and after disasters or major social disruptions as these network members may be geographically distant from survivors and therefore better situated to provide aid. For example, religious communities outside New Orleans in areas such as Baton Rouge and Biloxi immediately opened up shelters in the aftermath of Hurricane Katrina [25], Gotham and Powers [20]. Similarly, in the case of the opioid crisis, when individuals who struggle with substance use disorder are sent to detox and rehabilitation centers, their social ties become weaker, and connections with family members or close friends are hard to maintain. The sense of isolation experienced by individuals in the detox center has been studied in the past [26] and demonstrates the need for interventions that span the boundaries of family connections and involve "weaker" social connections.

The third and final type of connection, called linking social capital, is between a regular person and someone in power or authority. While a strong connection with family members and acquaintances are the most common, these may not be enough to guarantee individuals receive the support they need in case of emergency. Bridging and linking ties, while harder to create and maintain, can help vulnerable populations get ahead, especially when social ties become looser and geographical distance becomes a barrier that prevents vulnerable individuals to receive the emotional support they need.

The support availability of social and physical infrastructure available in a community might vary based on resources and socio-demographic factors. Individuals might have higher or lower access to emotional support, depending on where their family or friends currently reside, or instrumental support, determining their ability to find a stable residence or a job which in turn might impact individuals with substance use disorders [12].

While social ties to friends, family members, and individuals belonging to religious or civic organizations (bonding and bridging social capital) might provide immediate support in case of overdose or similar emergencies, connections to individuals in powers are likely to have a less immediate influence to the individual's recovery. Knowing someone who is well connected to decision-makers could have an impact on the opioid crisis over a longer period of time. The impact of government connections (linking social capital) might matter less in the short term, since political ties to the individual in power might not be enough to guarantee an immediate reduction in deaths.

The complexity of coordinating the efforts of various community stakeholders, including health workers, elected officials, policymakers, and business leaders, might suggest that government connections and political contributions are not strong enough determinants of an immediate impact on drug use. As Sampson and Wikstrom [17] found in their study of social capital in the Chicago neighborhoods, although residents seem to disengage and are more cynical in disadvantaged communities, community leaders become more intensely involved in seeking resources, often from afar. But this engagement requires integration and coordinated efforts which may lead to delayed responses and decreased efficacy [17].

Based on the literature mentioned above, we would expect fewer opioid incidents in areas with higher values of bonding and bridging social capital, while the association with the linking social capital could be less significant because of the complexity of factors that could impact opioid incidents.

Data Collection and Processing

In October 2018 and October 2019, we convened two Community Advisory Board meetings with individuals who live, work, pray, and/or have some other important connections to communities that are impacted by the opioid addiction in the city of Boston, Massachusetts. The discussion focused on how to make their communities safer from the damaging impact of addiction. Participants represented community health centers, large acute-care hospitals, faith-based organizations, community health workers, residents, institutes of higher education, frontline medical staff, and researchers. The advisory board noted that the Boston neighborhoods demonstrate great variance in recovery success, including in areas where high success rates might be expected owing to infrastructure support (e.g., a section of Boston commonly referred to as Methadone Mile/Recovery Road). It was recommended that data analysis focuses on the Boston neighborhoods to derive granular insights into social capital system dynamics and understand if the foundation of a community's resilience and recovery ability can also be explained through social support in addition to infrastructure. We, therefore, focused data collection and analysis on the city of Boston and its 23 diverse neighborhoods. Readily available datasets pertaining to opioid-related incidents and death are obtainable from the city's public health and safety departments.

Members of the focus group emphasized that the following factors may impact successful recovery: detox centers; transitional care (e.g., intensive outpatient); halfway houses; mutual help groups (Narcotics Anonymous, Alcoholics Anonymous, SMART recovery, faith-based groups, others); methadone, suboxone supplementation; and/or individual counseling. Adjunct treatment of comorbid psychiatric and medical illness are often an essential part of complete recovery as well; dreams and hopes for the future.

To represent bonding social capital, we used seven variables that measure the strong connections among individuals looking at similarities in demographic characteristics, attitudes, and resources. Bridging social capital was represented by seven variables that measure acquaintances or loose connections through ties that span social divisions and groups. And linking social capital was represented by six variables to measure connections of regular citizens to those in power.

We obtained data from four primary sources. We collected the approximate addresses of "sick assists" for the years 2015–2019 from the Boston Police Department. We then collected addresses of residents who died from fentanyl-related causes in 2015 from the Massachusetts Department of Public Health. We then relied on data from the Environmental Systems Research Institute (ESRI) to include pharmacies and community health centers that distribute opioid reversal drugs. Social capital data relevant to the study are also downloaded from ESRI's Community and Business Analyst sources at the block group level. The initial geodatabase includes a separate feature class for each variable (bonding, bridging, and linking). Table 18.1 illustrates the variables used in this study which are a subset of the ones used by Kyne and Aldrich [27]. Data associated with these variables are publicly available [28, 6, 29].

Results

The hot spot map in Fig. 18.1 identifies six areas of heightened opioid activity throughout the city of Boston. These Boston neighborhoods include East Boston, Downtown, the South End, Roxbury, Dorchester, and Jamaica Plain (Fig. 18.1). The area of highest density exists along Massachusetts Avenue, a well-known hotspot commonly referred to as the Methadone Mile, or Recovery Road. All areas demonstrate correspondingly high confidence levels confirming the clustered patterns are not random.

The spatial autocorrelation tool confirms clustering is significant where there is less than 1% likelihood that the clustered patterns identifying the six areas with

Table 18.1 The social capital variables used in this study, after Kyne and Aldrich [27]

	Social capital variable	Study variable	Source
Bonding			
1	Race similarity	Diversity index ranges from 0 (no racial diversity) to 100 (complete racial diversity)	(ESRI 2018CA) Block group
2	Educational equality	Absolute difference between % population with college education and % population with less than high school education	(ESRI 2018CA) Block group
3	Race/income equality	Gini coefficient ranges from 0 (perfect equality) to 1 (perfect inequality)	(US Census 2018AFF) Block group
4	Employment equality	Absolute difference between % employed and % unemployed civilian labor force	(ESRI 2018CA) Block group
5	Language competency	Calculated from the US Census Limited English Proficiency data set	(US Census 2018AFF) Block group
6	Communication capacity	% households with a telephone	(ESRI 2018CA) Block group
7	Non-elder population	% population below 65 years of age	(ESRI 2018CA) Block group
Bridging			
1	Religious organizations	Religious organizations per 10,000 persons	(ESRI 2018BA) Business and facilities search block group
2	Civic organizations	Civic organizations per 10,000 persons	(ESRI 2018BA) Business and facilities search block group
3	Social embeddedness—charitable ties	Member of charitable organizations (%)	(ESRI 2018CA) Block group
4	Social embeddedness—Church ties	Member of church boards (%)	(ESRI 2018CA) Block group
5	Social embeddedness—Fraternal ties	Member of fraternal orders (%)	(esri 2018ca) block group
6	Social embeddedness—Religious clubs	Member of religious clubs (%)	(ESRI 2018CA) Block group
7	Social embeddedness—Union ties	Member of unions (%)	(ESRI 2018CA) Block group
Linking			

(continued)

Table 18.1 (continued)

	Social capital variable	Study variable	Source
1	Political linkage	% voting-age population who are eligible for voting	(ESRI 2018CA) Block group
2	Local government linkage	% of local government employees working for local governments	(ESRI 2018CA) Block group
3	State government linkage	% of state employees working for the state governments	(ESRI 2018CA) Block group
4	Federal government linkage	% of federal employees working for federal agencies	(ESRI 2018CA) Block group
5	Political linkage—contribution	Contributed to political organization in the past 12 months (%)	(ESRI 2018CA) Block group
6	Social linkage—social services	Contributed to social services organization in the past 12 months (%)	(ESRI 2018CA) Block group
7	Religious linkage—religious contribution	Contributed to religious organization in the past 12 months (%)	(ESRI 2018CA) Block group
8	Political linkage—political activities	Attended political rally/speech/organized protest (%)	(ESRI 2018CA) Block group

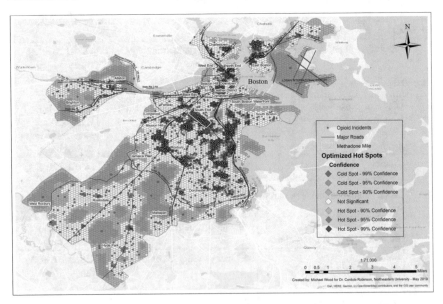

Fig. 18.1 Getis-Ord Gi* hotspot analysis of opioid incidents in Boston

Fig. 18.2 Spatial
autocorrelation results

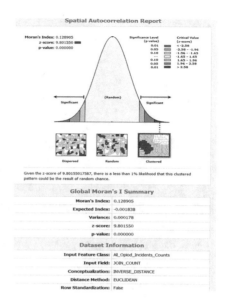

heightened opioid activity are the result of random chance. Figure 18.2 illustrates the results of the inverse distance spatial autocorrelation tool, which confirms that clustering is significant. All models indicate less than 1% likelihood that the clustered pattern is random.

Bonding Social Capital Map

To measure bonding, seven metrics were applied to reflect strong connections among individuals. Each of the seven variables serves as yield indicators to measure homophily and similarity. Specific bonding variables include: race similarity, educational equality, race/income equality, employment equality, percent population proficient English speakers, percent of households with a telephone, percent population below 65 years of age. At the block level, racial income similarity and gender income similarity metrics were unavailable and we could not include them in the social map. A social capital score of 0 is the lowest ranking and 1 the highest ranking. The data classification method is Equal Interval. The locations of opioid incidents including fentanyl death residence locations are plotted for context. Opioid incidents tend to occur in or immediately adjacent to the block groups with a medium bonding score. Opioid incidents occur in or immediately adjacent to medium bonding block groups (Fig. 18.3).

Opioid incidents appear to be concentrated along the Massachusetts Avenue corridor (Methadone Mile/Recovery Road), as well as Downtown and East Boston, where there are medium levels of bonding social capital. In the neighborhoods of

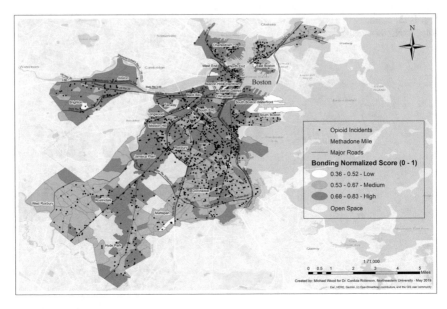

Fig. 18.3 Bonding social capital in Boston

Roxbury and Dorchester as well as East Boston, we observe medium levels of bonding social capital and large numbers of opioid incidents. We also observed that open space areas have very few to no opioid incidents.

Bridging Social Capital Map

Figure 18.4 visualizes bridging variables that reflect connections between individuals through ties that span social divisions and groups. The connections come from an individual's involvement in various civic organizations including religious organizations, charitable organizations, churches, fraternal orders, and unions. We were not able to collect data at the block level on the number of individuals affiliated with a religious organization per 10,000 persons, since the U.S. Census Bureau does not collect data on religious affiliation in its demographic surveys or decennial census (Public Law 94-521).

Similarly to the bonding social capital, a bridging score of 0 is the lowest ranking and 1 the highest. Equal interval classification divides the range of attribute values into equal-sized subranges with user-specified intervals. Class ranges are set to 3 and enable comparative visual interpretation. Central locations exhibit low connections; while peripheral locations in Jamaica Plain, Mattapan, and sometimes wealthier neighborhoods (Back Bay, Beacon Hill, and Downtown) show higher connections.

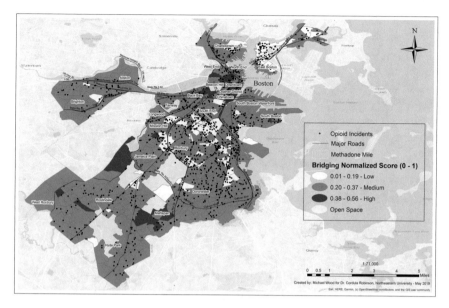

Fig. 18.4 Bridging variables social capital in Boston

Opioid incidents tend to occur in or immediately adjacent to the block groups with a low bridging score.

In neighborhoods where there is a higher concentration of people who are members of charitable organizations, members of church boards, members of fraternal orders, members of religious club, or members of union, there are fewer opioid incidents.

Data indicates that Roxbury and Dorchester, as well as the South End and East Boston neighborhoods have the most low bridging social capital block groups. Opioid incidents tend to occur in or immediately adjacent to the block groups with a low bridging score. Low bridging social capital appears to be concentrated along most of the Massachusetts Avenue corridor.

Linking Social Capital Map

Figure 18.5 examines the linking social capital variables measuring the connections of regular citizens to those in power. The variables include government connections, political contributions, social linkages by social services, religious linkages by contribution, and political links through openly political activities. All variables under Linking (a percentage) are normalized between 0 and 1 and equally weighted, and the data classification method is *Equal Interval*.

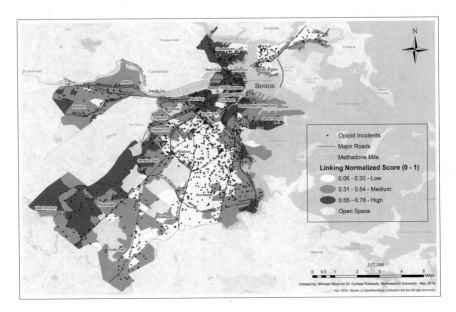

Fig. 18.5 Linking social capital in Boston

Central locations, such as the areas around Methadone Mile/Recovery Road, Roxbury, Dorchester, and Mattapan, exhibit low connections; while peripheral locations, and mostly wealthier neighborhoods (Back Bay, Beacon Hill, Downtown, and West Roslindale) show high connections to individuals in power. Opioid incidents correspond with a low linking social capital score. Results show that low linking social capital is more extensive throughout the Boston neighborhoods than bonding and bridging social capital. In particular, Roxbury and Dorchester, Mattapan, the South End, and East Boston include neighborhoods with low linking social capital block groups. In these block groups, we observe a higher number of opioid incidents though this doesn't dominate all groups with low scores (e.g., Mattapan).

Composite Map: Social Capital Score by Block Group

When overlaying all opioid incidents, there appears to be a strong visual correlation between low social capital scores and high numbers of opioid incidents. An exception to this observation is Downtown Boston/Chinatown that illustrates a cluster of incidents but have high social capital. Most low social capital block groups are in Roxbury and Dorchester, and East Boston neighborhoods. Fentanyl deaths/opioid incidents tend to concentrate more in or near block groups with a low social capital score, while fewer fentanyl-related deaths/opioid incidents occur in areas with high social capital scores.

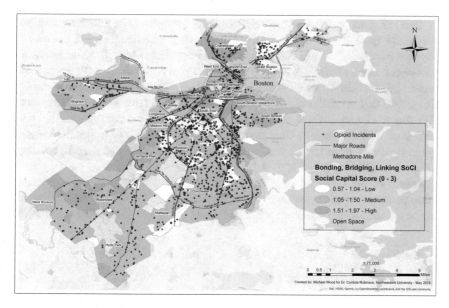

Fig. 18.6 Composite map of social capital: bonding, bridging, and linking

East Boston stands out, having both low social capital and high number of Fentanyl-related deaths. Also, the block groups in the vicinity of the Methadone Mile/Recovery Road are all low social capital block groups. Much of the Massachusetts Avenue corridor is comprised of low social capital block groups except for the Back Bay neighborhood near the Charles River (a fairly affluent sector of the city). Downtown, Beacon Hill, the Back Bay, Charlestown, and suburbs tend to have the highest social capital with fewer incidents. Figure 18.6 is a combination product of normalized scores calculated from the bonding, bridging, and linking variable datasets.

Regression Analysis

To quantitatively test the visual appraisal, we conducted a Geographically Weighted Regression (GWR) to examine whether social capital factors explain opioid-related incidences. Social capital variables are the independent variables and the number of opioid incidences is the dependent variable. We use the social capital variables to predict where we might see opioid incidents. To ensure our models are not biased, we conducted spatial autocorrelation for all residuals, and we found that they were random, not clustered, thus significant. Areas demonstrating more incidents than expected, might show other factors are in play, e.g., mobility within the region; similarly for those areas with cold spots.

With reference to the specific neighborhood, Downtown Boston, Roxbury, and Dorchester have the highest residuals. Areas such as Mattapan and pockets of Brighton have low incidents compared to what we would expect. The Massachusetts Avenue corridor and Methadone Mile/Recovery Road is prominent as there are many more incidents than we would expect based on the bridging social capital variables. This means that in these areas there might be other factors impacting the occurrence of deaths or sick assists that could not be explained with the connections that residents have with religious or civic organizations. Figure 18.7 illustrates the maps of opioid incidents and what we would have expected given the observed to bonding, bridging, and linking social capital.

Results of the GWR analysis suggest that bridging social capital has a strong influence in terms of correlation with opioid incidents. Results on bonding social capital are inconclusive, neither confirming nor refuting the initial observation. The

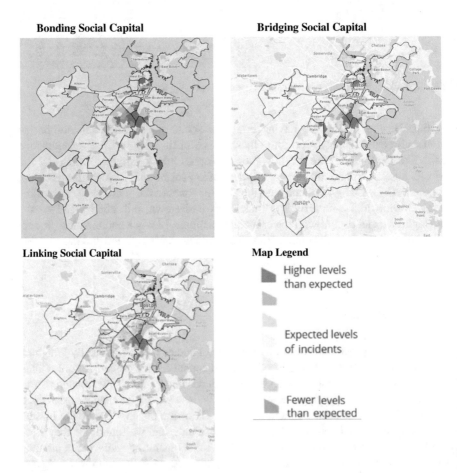

Fig. 18.7 Maps of bonding, bridging, and linking social capital: expected data and GWR results

GWR results confirm the vague correlations of high occurrences of opioid incidents and block groups with low linking social capital. Linking social capital is extensive throughout the Boston neighborhoods, although it does not dominate all groups with low scores (e.g., Mattapan).

Discussion

Our findings suggest that in the Boston neighborhoods where social capital is low, there is a moderately high incidence of opioid deaths and sick assist calls, as we observed more hot spots and fewer cold spots. In particular, we found that opioid-related incidents will go down in areas where bonding and bridging are at a medium to high level. The hot spot map identifies six areas of heightened opioid activity throughout the city of Boston, where bridging social capital variables are higher. Bridging social capital shows the highest influence in terms of correlation with opioid incidents. Our analysis shows that in neighborhoods where residents are involved in charitable organizations, where people gather around religious organizations, or where unions are more active, people help each other more and might be aware of actions to take to prevent opioid-related deaths. Individuals living in a neighborhood where there are more churches or civic organizations might be more likely to receive support when they are struggling because of the community propensity to help and be more outward focused.

Charitable or religious organizations might act as catalysts for building social support, as educators, as institutions where people gather to find solutions to community and individual problems, or where people whose strong ties have deteriorated because of geographic distance, can find support and first aid. Given the importance of the bridging social capital, we would recommend the development of initiatives and incentives to build synergies among civic and religious organizations that can act as an immediate social support system when strong ties (i.e., bonding social capital) become unavailable.

In certain areas in Boston, such as Methadone Mile/Recovery Road, Downtown Boston, and East Boston, we see a high concentration of opioid incidents, as well as medium levels of bonding social capital, and low levels of bridging social capital. In these areas residents share fewer socio-demographics characteristics, and seem to be less involved in charitable, religious, or civic organizations. As demonstrated by studies in the African American and Latino communities [30], church-based interventions have the potential to yield stigma reduction. Fighting the opioid crisis requires a community-based approach, by developing partnerships with local institutions and organizations and investing in opportunities to support individuals who have lost their family/friends/connections. Tailoring interventions to single race-ethnic groups may not be the best approach in diverse community settings, which are common among various Boston neighborhoods.

The role played by the bridging social capital can also be interpreted by looking at the work done by Community Health Workers (CHWs). According to the Community Health Workers in Massachusetts report (2009), 90% of CHWs report working with individuals with substance use disorders. This has prompted elected officials, policymakers, and business leaders to engage and partner with community health centers to more directly tackle the opioid crisis on the premise that every individual struggling with substance misuse should have equitable access to treatment. CHWs provide direct services, such as informal counseling, social support, care coordination, and health screenings which help to alleviate the stigma associated with the use of drugs, often a barrier to care and treatment [31].

Results indicate that bonding social capital is not directly correlated with higher incidences of sick assists or opioid-related deaths. This means that neighborhoods where residents share common socio-demographic characteristics, income, language, and age, are not necessarily going to have lower cases of opioid incidents. As we know, individuals from a wide range of income, net worth, educational attainment, occupation, and race have all been impacted by the epidemic. Future studies could focus on further exploring these demographics to better understand the community and their neighborhoods, with the goal of targeting similar demographics with education and community outreach that could reduce stigma and save lives.

Findings suggest that a community's resilience to opioid crisis can be explained by social composition and support (or lack therefore) as opposed to critical infrastructure. Our findings show that recovery infrastructure are pervasive within high incident areas and almost follows such locations. This indicates that a community's resilience to the opioid crisis can be measured by its level of social capital, by differentiating between bonding, bridging, and linking social capital.

References

1. E.F. McCance-Katz, P. George, N.A. Scott, R. Dollase, A.R. Tunkel, J. McDonald, Access to treatment for opioid use disorders: medical student preparation. Am. J. Addict. (2017). https://doi.org/10.1111/ajad.12550
2. F.G. Robinson, C. Lee, E. Avalon, P. Edmonds, MA Opioid crisis: spatial data-driven approaches for community resilience. GIS-Pro & CalGIS (2018)
3. B. Bearnot, J.F. Pearson, J.A. Rodriguez, Using publicly available data to understand the opioid overdose epidemic: Geospatial distribution of discarded needles in boston, Massachusetts. Am. J. Public Health (2018). https://doi.org/10.2105/AJPH.2018.304583
4. T.J. Stopka, M.A. Goulart, D.J. Meyers, M. Hutcheson, K. Barton, S. Onofrey, K.K.H. Chui, Identifying and characterizing hepatitis C virus hotspots in Massachusetts: a spatial epidemiological approach. BMC Infect. Dis. (2017). https://doi.org/10.1186/s12879-017-2400-2
5. S.M. Kelly, K.E. O'Grady, R.P. Schwartz, J.A. Peterson, M.E. Wilson, B.S. Brown, The relationship of social support to treatment entry and engagement: the community assessment inventory. Subst. Abuse (2010). https://doi.org/10.1080/08897070903442640
6. F.H. Norris, S.P. Stevens, B. Pfefferbaum, K.F. Wyche, R.L. Pfefferbaum, Community resilience as a metaphor, theory, set of capacities, and strategy for disaster readiness. Am. J. Commun. Psychol. (2008). https://doi.org/10.1007/s10464-007-9156-6

7. D.P. Aldrich, *Building Resilience: Social Capital in Post-disaster Recovery* (University of Chicago Press, 2012)
8. D.P. Aldrich, M.A. Meyer, Social capital and community resilience. Am. Behav. Sci. (2015). https://doi.org/10.1177/0002764214550299
9. J.S. Coleman, S.J. Coleman, Social capital in the creation of human capital. Am. J. Sociol. (1988). https://doi.org/10.1086/228943
10. D. Panebianco, O. Gallupe, P.J. Carrington, I. Colozzi, Personal support networks, social capital, and risk of relapse among individuals treated for substance use issues. Int. J. Drug Policy (2016). https://doi.org/10.1016/j.drugpo.2015.09.009
11. Y.W. Cheung, N.W.T. Cheung, Social capital and risk level of posttreatment drug use: implications for harm reduction among male treated addicts in Hhong Kong. Addict. Res. Theory (2003). https://doi.org/10.1080/1606635021000021511
12. S.T. Ennett, S.L. Bailey, E.B. Federman, Social network characteristics associated with risky behaviors among runaway and homeless youth. J. Health Soc. Behav. (1999). https://doi.org/10.2307/2676379
13. D.N. Nurco, T.W. Kinlock, K.E. O'Grady, T.E. Hanlon, Early family adversity as a precursor to narcotic addiction. Drug Alcohol Depend. (1996). https://doi.org/10.1016/S0376-8716(96)01299-9
14. M.D. Burkey, Y.A. Kim, W.R. Breakey, The role of social ties in recovery in a population of homeless substance abusers. Addict. Disord. Treat. (2011). https://doi.org/10.1097/ADT.0b013e3181ea7511
15. C.A. Latkin, W. Hua, K. Tobin, Social network correlates of self-reported non-fatal overdose. Drug Alcohol Depend. (2004). https://doi.org/10.1016/j.drugalcdep.2003.09.005
16. K.M. Broome, D.K. Knight, K. Knight, M.L. Hiller, D.D. Simpson, Peer, family, and motivational influences on drug treatment process and recidivism for probationers. J. Clin. Psychol. (1997). https://doi.org/10.1002/(SICI)1097-4679(199706)53:4%3c387:AID-JCLP12%3e3.0.CO;2-C
17. R.J. Sampson, P.O.H. Wikström, The social order of violence in Chicago and Stockholm neighborhoods: a comparative inquiry. Order Confl. Violence (2008). https://doi.org/10.1017/CBO9780511755903.006
18. R.J. Sampson, C. Graif, Neighborhood social capital as differential social organization. Am. Behav. Sci. (2009). https://doi.org/10.1177/0002764209331527
19. S. Galea, A. Nandi, D. Vlahov, The social epidemiology of substance use. Epidemiol. Rev. (2004). https://doi.org/10.1093/epirev/mxh007
20. K.F. Gotham, B. Powers, Building resilience: social capital in post-disaster recovery. Contemp. Sociol. J. Rev. (2015). https://doi.org/10.1177/0094306114562201a
21. L.J. Hanifan, The rural school community center. Ann. Am. Acad. Polit. Soc. Sci. (1916). https://doi.org/10.1177/000271621606700118
22. M. McPherson, L. Smith-Lovin, J.M. Cook, Birds of a feather: homophily in social networks. Ann. Rev. Sociol. (2002). https://doi.org/10.1146/annurev.soc.27.1.415
23. E. Chamlee-Wright, The cultural and political economy of recovery: Social learning in a post-disaster environment, in *The Cultural and Political Economy of Recovery: Social Learning in a Post-Disaster Environment* (2010). https://doi.org/10.4324/9780203855928
24. M.S. Granovetter, The strength of weak ties. Am. J. Sociol. (1973). https://doi.org/10.1086/225469
25. D.P. Aldrich, The power of people: social capital's role in recovery from the 1995 Kobe earthquake. Nat. Hazards (2011). https://doi.org/10.1007/s11069-010-9577-7
26. T. Orzeck, A. Rokach, Men who abuse drugs and their experience of loneliness. Eur. Psychol. (2004). https://doi.org/10.1027/1016-9040.9.3.163
27. D. Kyne, D.P. Aldrich, A New Framework for Capturing Social Capital, Working Paper (2018)
28. B.H. Morrow, *Community resilience: a social justice perspective*. CARRI Research Report (2008)
29. K.J. Tierney, M.K. Lindell, R.W. Perry, J.H. Press, Facing the unexpected: disaster preparedness and response in the United States (Google eBook). Nat. Hazards Dis. (2001). https://doi.org/10.17226/9834

30. K.P. Derose, L.M. Bogart, D.E. Kanouse, A. Felton, D.O. Collins, M.A. Mata, M.V. Williams, An intervention to reduce HIV-related stigma in partnership with African American and Latino churches. AIDS Educ. Prev. (2014). https://doi.org/10.1521/aeap.2014.26.1.28
31. L. McFarling, M. D'Angelo, M. Drain, D.A. Gibbs, K.L.R. Olmsted, Stigma as a barrier to substance abuse and mental health treatment. Mil. Psychol. (2011). https://doi.org/10.1080/08995605.2011.534397

Chapter 19
Reward-Based Crowdfunding as a Tool to Constitute and Develop Collaborative Innovation Networks (COINs)

Michael Beier and Sebastian Früh

Abstract The concept of "Collaborative Innovation Networks" (COINs) has been successfully applied in many projects over the past 15 years to detect COINs in given situations and to enhance the behavior of related actors in the corresponding social networks. However, what might be missing is an easily applicable tool, which helps potential initiators of an innovative endeavor in a guided process to initially constitute and further develop a COIN over several stages. In this paper, we follow the idea that reward-based crowdfunding campaigns could be such a practical tool. Therefore, we develop a conceptual framework of how reward-based crowdfunding can be applied to support the constitution and development of COINs.

Introduction

Most people perceive reward-based crowdfunding mainly as a relatively new way to raise funding via the Internet. However, reward-based crowdfunding can be much more than that, especially when it is applied to promote specific purposes of collaborative innovation. It provides a practicable tool to constitute and develop collaborative innovation networks (COINs) that can be applied as a complement to the classical COINs approach. In this paper, we develop based on an extensive literature review a conceptual framework of how reward-based crowdfunding can be applied to support the constitution and development of COINs.

The paper is structured as follows: in the next section, we briefly introduce relevant theoretical foundations and the relevant state of research on collaborative innovation networks (COINs) in the context of this paper. In the third section, we do the same for reward-based crowdfunding. In section four, we combine both approaches and

M. Beier (✉) · S. Früh
University of Applied Sciences of the Grisons, Swiss Institute for Entrepreneurship, Chur, Switzerland
e-mail: michael.beier@fhgr.ch

S. Früh
e-mail: sebastian.frueh@fhgr.ch

© Springer Nature Switzerland AG 2020
A. Przegalinska et al. (eds.), *Digital Transformation of Collaboration*,
Springer Proceedings in Complexity,
https://doi.org/10.1007/978-3-030-48993-9_19

present our framework on how reward-based crowdfunding can be applied as a tool to constitute and develop collaborative innovation networks. Finally, we present our conclusions and implications for practical applications and subsequent research.

Collaborative Innovation Networks (COINs)

Collaborative innovation networks (COINs) can basically be defined as "virtual communities interacting on a global scale …˙ made up of self-motivated people who share a common vision, meeting on the web to exchange ideas, knowledge, experiences and to work in a collaborative way to achieve a common goal" [18], p. 6. Correspondingly, COINs are dynamic social networks, in which actors innovate, collaborate, and communicate on new ideas [25]. The main components of their definition are: (1) an evolving community of likeminded people, (2) connected and interacting to a significant extent via digital channels, (3) to commonly pursue to develop and disseminate some kind of innovation [29].

The development of COINs can generally be structured into four phases [27]. First, the network initially originates from an individual initiator or an informal group, which starts an innovative endeavor. Second, a team constitutes around the initiator(s) or the idea. Such a core team is called a "Collaborative Innovation Network (COIN)". It builds the core origin of the innovative endeavor and consists of 3–15 people. Third, after further developing the idea within the core team more likeminded people have to be attracted to the social movement to support the further development of the innovation as well as its dissemination. This stage is called "Collaborative Learning Network (CLN)". Corresponding networks consist of hundreds of people who actively exchange knowledge on regarding topics in the context of the innovation, and therefore directly contribute to the development with regard to the content of the collaborative endeavor. In the fourth and final phase of dissemination, the collaborative innovation network needs to reach even more people, most of them only in the role of consumers of relevant information as well as services or products provided on the basis of the new innovation. This stage is called "Collaborative Interest Network (CIN)" and often consists of thousands of people who use or further spread an innovation. In addition to the successive phases of COIN development described in this section, the four different roles in the COIN network (initiator(s), COIN core team, CLN, and CIN) can also be applied to locate or analyze specific actor types in fully established and extended COIN networks after the fourth phase has been completed.

However, the COINs concept and the constituting social networks of its developmental stages are not directly perceivable nor does the concept provide hands-on tools for people to originally constitute (and develop) such networks on their own. Rather the general COINs approach applies a scientific consulting perspective (mainly based on social network analysis methods) to help optimize already existing social networks for collaborative innovation purposes by evidence-based advice for social network development and virtual mirroring [26]. In this context, virtual mirroring means that specific characteristics of COIN networks are visualized or computed and communicated to respective actors in the network or organization under investigation to provide them a "virtual mirror". This mirroring allows these actors to better reflect on their individual and collective interaction patterns and to adapt their behavior towards more beneficial patterns to enhance specific objectives of their collective, like increases in innovation output or customer satisfaction [28]. Respective characteristics to detect COINs are called "honest signals" and cover measures of the structure, dynamics, and content of interactions in social networks [27].

The COINs approach has been successfully applied in many projects over the past 15 years to detect COINs in organizations and to adapt the behavior of related actors in the corresponding social networks [26–28]. However, at present the COINs framework is more reactive in the sense that it mainly helps to analyze structures of given social networks, detect COINs (honest signals) in it, or to provide recommendations and tools to adapt the behavior of actors in the networks (in particular social network analysis, Condor software, virtual mirroring). What is missing, is an easy applicable tool, which helps potential initiators of an innovative endeavor in a guided process to constitute and further develop a COIN over several stages, even without being educated in social network analysis (even though it would be beneficial for them to be). In this paper, we follow and apply the idea that reward-based crowdfunding campaigns could be such a practical tool.

Reward-Based Crowdfunding (RBCF)

Crowdfunding is mainly known as a new way of digital fundraising for innovative projects and ventures, which has become popular over the past years. Instead of asking banks, venture capitalists, business angels, or governmental agencies for funding, project initiators make a direct call via the Internet to a virtually related "crowd" of potential stakeholders for financial support for their endeavor [47]. Generally, crowdfunding can be defined as an "open call, mostly through the Internet, for the provision of financial resources either in the form of donation or in exchange for the future product or some form of reward to support initiatives for specific purposes" [10], p. 588). Supporting this, crowdfunding platforms act in this regard as digital intermediaries in two-sided markets matching fundraisers and funders via crowdfunding campaigns operated in the platform [11].

Such crowdfunding campaigns can apply one of two different campaign models: First, in "all-or-nothing" campaigns project initiators define a certain amount of money they ultimately need to raise via the campaign to be finally able to realize their intended project (the "funding goal"). The amount of money raised on the platform during an all-or-nothing campaign is only paid out to the project initiators if the amount reaches at least this defined threshold. Otherwise after the end of the campaign all pledges are paid back to the backers [7, 34, 40]. Second, crowdfunding campaigns can be run on the basis of a "keep-it-all" model. In campaigns of this type, the project initiators get the amount raised during a campaign in any case independently of any threshold [17, 31].

Crowdfunding research differentiates between four specific crowdfunding types in dependence of the kind of goods project initiators offer in exchange for funding they receive (e.g., [2, 9, 11, 16]): Initiators can offer equity shares (equity-based crowdfunding), particular interest rates (lending-based crowdfunding), some kind of service or product (reward-based crowdfunding) as well as some kind of activities to achieve a mutually desired or pro-social goal (donation-based crowdfunding) in exchange for funding. In this paper, we focus only on reward-based crowdfunding (hereinafter referred to as RBCF).

RBCF differs to all other crowdfunding types, as it is the only one where funders can receive a product or service in exchange for their financial support. Therefore, mostly funding decisions in RBCF campaigns are less decisions of investors but more decisions of customers "buying" an innovative product or service as a reward. In this regard, RBCF combines elements from social media and e-commerce [8, 9]. On the one hand, RBCF platforms are specific social media platforms where campaign initiators can present their project by simply uploading texts, photos, or videos [38, 45, 56]. Furthermore, crowdfunding platforms offer functionalities of online social networks where project initiators maintain an own profile for their project and can post updates, which are communicated to followers of the project on the platform or via other digital channels [9, 40, 45, 57]. On the other hand, campaign pages in RBCF platforms provide projects a web-based point of sale where they can offer their services or products (mainly as "rewards", but also as the overall project outcomes) to potential supporters in the role of customers. Similar to specific product pages in conventional online shops, therefore, campaign initiators have to generate traffic to their individual project page on the RBCF platform and convert it there to purchases in their campaign [8, 43].

Whereas the keep-it-all model makes it easier for project initiators in RBCF campaigns to generate some funding for their project, the all-or-nothing model forces project initiators to further plan their project and to estimate the financial resources they need [7]. In addition, they should make additional calculations about their network of supporters and potential customers, and define what exactly they have to offer them in exchange for their contributions [6]. From an entrepreneurial perspective, an RBCF campaign is a good tool to bring a funding team in an experimental setting together [3]. The all-or-nothing model, in addition, allows the team to run a proof-of-concept. On the one hand, this includes internal questions of the team (What exactly do we want to reach in the long term? What could be a feasible but goal-oriented project as a first step in our long-term endeavor? Do we have all resources and capabilities for this project?). On the other hand, this refers to questions regarding external stakeholders (Is our idea interesting and relevant for a sufficient number of external stakeholders? Are we able to reach and attract these stakeholders in a certain period of time and with our given resource endowment? Are we able to operate on all communication channels necessary to reach our target groups?).

In the first years, practitioners applying RBCF for their projects as well as scientific research analyzing them strongly emphasized the financial fundraising results as the main objectives of RBCF campaigns. Therefore, most studies so far mainly addressed in their tested hypotheses factors influencing funding success of RBCF campaigns, especially in terms of overall successful funding (all-or-nothing) or total amounts raised in campaigns (e.g., [9, 33, 35–37, 42]). However, in recent years some scientists began to perceive RBCF also as a collaborative innovation tool and therefore to explore what additional benefits RBCF campaigns provide to project initiators, in terms of community building, co-creation, crowdsourcing, market knowledge, open innovation, or the development of shared social identities (e.g., [5, 6, 14, 20, 23, 39, 50, 55]). However, until now this perspective has been applied only in part by a few project initiators in practical RBCF applications. The main objective for most initiators in practice is still the funding they can raise with their RBCF campaign [5, 14].

In many cases, running an RBCF campaign is just a first step of innovative projects. For instance, analyses of Kickstarter data showed that more than 90% of successfully financed projects went on with their endeavor after their campaign. 32% started ventures, which generated more than 100,000 USD yearly revenues and on average 2.2 jobs per successful project [41]. On the other hand, initiators of failed RBCF campaigns sometimes use their early learnings during the campaign to further develop their project or to change the central characteristics of the project to better meet the expectations of their future market. They just draw their conclusions from their experiences during the campaign and enter the market with their finalized product [55]. Furthermore, (more risk-averse) project initiators, which failed in an RBCF campaign, sometimes try another campaign to proof their adapted concept again, before they go on, finalize their product, and enter the market [30, 39].

Besides the development of the innovation itself, RBCF campaigns also can help innovative teams to further develop their community around their project [23]. One important activity of initiators preparing and running an RBCF campaign is to systematically reach out for relevant stakeholder groups, especially via digital channels [6, 24, 40, 53]. In addition to realizing the full potential of all relevant stakeholder groups, some project teams also make sure that their RBCF campaign generates them additional access to new people outside their current network for the further development of their project [5]. For instance, some teams design their whole campaign suchlike that it generates as many email addresses of (new) supporters as possible. Other teams use social rewards in their RBCF campaign (especially events or event series to physically meet their stakeholders and supporters) to build a vibrant community around their project [5]. In general, many RBCF campaigns find individual ways to tie stakeholders, which have been useful and supportive during the campaign, to their project to be accessible for future activities [19, 20].

RBCF for COINs: A Conceptual Framework

The descriptions above on developments in collaborative innovation networks (COINs), of the specific phases, and on teams preparing and running reward-based crowdfunding (RBCF) campaigns show manifold similarities between both approaches, particularly in regard to socio-cognitive developments, social network dynamics, and activities over different stages of project development. Whereas the COINs concept is more a science-based consulting approach to analyze and optimize given innovation networks, RBCF is a practical tool to constitute innovation teams and to promote their (digital) interactions with large networks ("crowds") based on a predefined process. Therefore, it seems valuable to combine both approaches by applying the established logics, process models, and success factors of RBCF to support the constitution and development of COINs.

Our respective conceptual framework of COIN development by application of RBCF campaigns is structured in two phases: first, a "set-up phase" where an idea is originally generated and a COIN core team is constituted; second, a "scale-up phase" where the core team of the COIN can be further extended with a CLN and a CIN.

Set-Up Phase

In the set-up phase, one or more initiators have an idea for an innovative endeavor and decide to start a project. In the COINs concept, this means that a close network of intrinsic motivated collaborators has to be constituted [27]. However, the COINs framework itself does not provide concrete activities or a predefined process to set up such a COIN. It is more focused on how to detect already existing COINs in social networks and how to moderate further network development to improve their innovation outcomes. Therefore, potential initiators of a COIN face the challenge of how they can initially constitute their COIN. Especially, in a corporate context (in and between companies or other organizations), it seems difficult to set up a new innovative movement out of informal activities in a diffuse setting of expectations, requirements, and capacities of potential co-initiators. Regarding the constitution of a COIN, the application of an RBCF campaign provides several advantages (see Table 19.1).

Initiators: First of all, an RBCF campaign provides a concrete *starting point* for an innovative endeavor with a project character. It may be difficult to constitute a COIN by starting an informal gestation process within or between organizations and to organize a team around a more or less diffuse idea for an innovative endeavor. An RBCF campaign as a concrete starting point might help to overcome such obstacles.

In this regard, bringing together a group of likeminded people for a more focused RBCF campaign also provides a *concrete project scope* for all participants, defining a first step for the potentially following endeavor. On the one hand, potential collaborators know more in detail what the campaign will be about. On the other hand, the

Table 19.1 Set-up phase of RBCF application for COIN constitution

Initiator(s)
Starting point with project character
Concrete project scope/calculable commitment
Reduced need for initial project budgets
Experimental setting
COIN core team
Team constitution
Shared social identity
Segmentation of external stakeholders
Coordinate/balance expectations and commitments
First level proof

RBCF campaign is limited and transparent in scope so that the group members only have to give a more *calculable commitment* (in time, resources, and workload) to a first step of the collaborative endeavor. Many examples of COINs processes have been observed in student projects during one semester, which have a similar scope in time and effort [26].

Even though fundraising aspects are not in the center of this framework, the by-effect of raising funding for the next step of an innovative project by running an RBCF campaign also helps to engage potential team members. The generation of funding (budget or other organizational resources) of innovative endeavors is often the main problem for informal innovation teams in and between organizations [15, 21]. As an RBCF campaign per se includes the generation of funding for the next steps of an innovative endeavor, it *reduces the need for initial project budgets* at the early beginning (for instance from members of the top management or other organizational sponsors). However, a successfully executed RBCF campaign might also help to get additional funding from organizational parties or other external sources [52].

In addition, according to the basic ideas of the Lean Start-up concept, an RBCF campaign allows an innovation team to test some fundamental assumptions of their approach in an *experimental setting* [3, 13]. As we will see in the following descriptions, an RBCF campaign also helps an innovation team to find out in a preliminary project stage if it really fits together and if it obtains all relevant capabilities and resources to advance the project beyond a certain level.

COIN Core Team: The core team of a COIN is essential for the further development of the innovation project. The *team constitution* for an RBCF campaign helps to identify missing capabilities and resources within the team. Similar to COINs, in RBCF a team of campaign initiators needs to combine different capabilities in regard of knowledge about the innovation itself, operational and management resources, as well as an adequate level of social capital to successfully run their campaign [1, 24, 32, 39, 40, 49, 53]. On the one hand, to successfully develop and spread an innovation it needs an adequate network of supporters well connected to the team members; on the other hand it needs the ability of the team members to reach and activate these contacts via digital and social media channels [27]. Furthermore, well-coordinated team processes are essential for campaign development and preparation in RBCF [5].

A further challenge during the set-up phase, where the team develops and prepares its RBCF campaign, is to develop a shared "project story" in the team [5, 6]. On the one hand, this means that the team really has to work out in detail what they plan to do in their intended project, before they actually start to realize it. On the other hand, they have to develop a *shared social identity* within the team [20]. This process seems quite similar to the one in the COINs concept, where a team has to develop collective consciousness between each other as well as with relevant members of their stakeholder network [26, 27]. However, the COINs concept provides only communication and interaction related advice on how such collective consciousness can be fostered during the constitution of a COIN. In contrast, the planning and preparation of an RBCF campaign is more concrete and directly applicable during the team constitution. It provides a clear setting and a certain pressure for a project team to storm and norm the team building in the form of their first milestone of starting and running a successful RBCF campaign [54].

However, it is not enough that the project story meets the internal requirements and expectations of all team members. In addition, it needs an adequate *segmentation of external stakeholders* (in particular potential supporters, customers, and communicators) to prepare a successful RBCF campaign [5]. Therefore, it is necessary that the team anticipates the expectations of specific segments of all stakeholder segments and their social contacts to be able to activate them to support their endeavor [4, 48]. On the one hand, this regards to the financial support of the campaign. On the other hand, this regards to the motivation for all contacts to spread the story of the innovation project in their community [20]. Against this background, for successful RBCF campaigns it is essential for the initiator team to develop a sophisticated segmentation of relevant stakeholder groups, including support motives, reward expectations, willingness and ability to pay, online and offline channels to reach them as well as the size of each segment [46]. However, similar considerations in the early stages of an innovative endeavor are also essential during the constitution of a COIN [27]. Therefore, also for the further development of COINs, it seems relevant to anticipate early the expectations and motivations of relevant stakeholders outside the core team to be prepared for later stages of their network development (CLNs and CINs).

The most common RBCF model of "all-or-nothing" also forces the team members in this early stage to discuss intensely and decide about a realistic scope of their project to define an adequate funding goal for their campaign [7, 40]. On the one hand, the funding goal should be high enough to really allow the team to successfully realize the project in the intended scope. On the other hand, very high funding goals compared to the planned activities in the project make a bad impression on potential supporters [6]. Furthermore, the higher the funding goal is the higher the risk to financially fail becomes in the RBCF campaign [9, 40]. Therefore, the collaborative discussion and decision on the funding goal for the campaign is also a good occasion to *coordinate and balance the expectations and commitments* between the members of the innovation team.

Finally, the preparation of an RBCF campaign provides an internal *first-level proof* that the team is endowed with an adequate level of capabilities and resources to at least start a campaign. Eventually given weaknesses regarding resource endowment or low process qualities within the team can be identified in an early stage and failure becomes obvious before the actual innovation project has started. For instance, platform data show that a significant percentage of teams, which already had started to enter their campaign data into the crowdfunding platform, actually failed to reach the point where they really published the campaign on the platform. All these teams got right at the beginning of their collaborative endeavor a valid and clear signal that they were missing some essential resource or capability in their core team to succeed in that constellation. However, an RBCF campaign also allows the realization of learning processes by the team during more than one campaign (or campaign preparation) potentially leading to a successful campaign at the end [39].

Scale-up Phase

After the invention and the constitution of a core team (during the set-up phase), in the "scale-up phase" the collaborative innovation network has to be extended to reach further knowledge providers ("Collaborative Learning Networks", CLN) as well as users, customers, and other supporters who might be helpful to spread the innovation ("Collaborative Interest Networks", CIN) (see Table 19.2). Within the COINs approach, RBCF campaigns are already mentioned as a valuable tool for these purposes [27]. However, concrete applications have not been specified so far.

Collaborative Learning Network (CLN): On the one hand, an RBCF campaign allows the core team to find out to what extent it is able to attract relevant stakeholders for the further *development* of the innovation. This is in line with the idea of CLNs in the COINs concept and regards mainly to the further technical development of the solution as well as business collaborations, which help to enhance the quality of

Table 19.2 Scale-up phase of RBCF application for COIN development

Collaborative learning network (CLN)
Development
Technology development partners
Business development partners
Collaborative interest network (CIN)
Customers
Pre-Market check
Prosumers
Communicators
Communication partners
Digital word-of-mouth

the finally resulting innovation. For instance, RBCF campaigns can help to establish new partnerships with companies and organizations providing complementary knowledge, technologies, or services to the own innovative solution [5, 32]. Also during the development and preparation of the RBCF campaign, potential partners can be approached to define and offer collaboratively a reward for the campaign. Therefore, RBCF campaigns provide a good opportunity to get in touch with new partners and to test with them potential fields of collaboration [5]. In addition, during their running RBCF campaign initiator teams also are able to identify specific individuals or informal groups in their environment, which obtain knowledge or capabilities relevant for the further development of the project and to integrate them into their CLN. For many active supporters of RBCF campaigns, for instance, to get in touch with the people behind the project is one of the central motivations for their engagement [22].

Collaborative Interest Network (CIN): On the other hand, RBCF campaigns can help a core team of a COIN to better estimate the extent of interest for their resulting product or service in the market [27]. Therefore, another group of relevant external stakeholders perceives an RBCF campaign more from a *customer* perspective. This is in line with the idea of CINs in the basic COINs concept, where actors' main interest in an innovation lays in their own usage of it [25]. Basically, RBCF campaigns provide a "pre-market check" of the general interest for the topic or the concrete demand for specific solutions of an innovation [6, 55]. More concrete RBCF campaigns also provide manifold information on preferences, buying intentions, and willingness to pay about potential customers [1, 14, 32]. During an RBCF campaign many stakeholders offer ideas or communicate their preferences, for instance, in the comments of the campaign page on the crowdfunding platform, in online communities (e.g., communities on upcoming gadgets, new technologies, or environmental topics), or in social media communications about the running campaign [1]. In addition, the portfolio of rewards as well as specific customizable rewards often offers supporters various opportunities to engage as "prosumers" in activities of collaborative co-design [44]. Detailed information on user preferences and customer needs, however, can be derived to a much further extent, if an RBCF campaign has been especially designed as an experiment. For instance, a promising opportunity to generate information about customers' needs and willingness to pay in the target market is the application of specially designed rewards (including a respective set of choice options and according prices) [5]. In this regard, the experimental logic of the Lean Start-up approach can be applied to that extent, that a COIN team designs its RBCF campaign specially to test central hypotheses for their intended innovative endeavor [3, 13].

Another set of relevant stakeholders in line with the CINs in the COINs concept are actors mainly in the role of *communicators*, aggregating and transferring information and messages about an innovative endeavor, a running RBCF campaign, or a new product or service. In this regard, an adequate preparation for an RBCF campaign also includes a systematical acquisition of (potential) communication partners for the campaign. Therefore, an initiator team seeks media coverage or collaborations with other communication partners like cross postings in company newsletters or posts of relevant influencers in the field [5]. Such activities help to extend the reach of the core team to better spread the innovation. Another set of relevant activities to prepare and run a successful RBCF campaign is the promotion of (digital) word-of-mouth [12, 51, 53]. On the one hand, this means that initiator teams need to establish adequate presence in relevant social media platforms. On the other hand, in addition, the team has to provide a continuous news stream to keep the communication waves in their own digital channels as well as of their connected communicators going [36].

Conclusions and Implications

In this paper, we developed and presented a conceptual framework of how RBCF can be applied to support the constitution and development of COINs through four different stages. With this framework, we intend to inspire innovators and innovation teams to perceive RBCF as more than just another opportunity for fundraising. As described in this paper, an RBCF campaign can be a helpful and practical tool to set up and scale-up collaborative innovation networks. Therefore, RBCF campaigns can be applied complementarily to the scientific consulting approach of the original COINs concept. However, in the opposite direction practical applications of RBCF campaigns also could benefit from further usage of social network analyses in line with the COINs concept. Correspondingly, there are three kinds of applications of the concept presented in this paper: First, RBCF campaigns can be applied to set up and scale-up collaborative innovation networks. Second, in the opposite direction, the tools and metrics of the COINs concept can be applied to optimize the network development and utilization of RBCF campaigns. Third, innovative endeavors might generally benefit most of a combined application of RBCF campaigns and the COINs approach.

Figure 19.1 provides an integrated model on how RBCF and the COINs approach might be applied together to foster the constitution and development of collaborative innovation endeavors. In this regard, both approaches complement each other

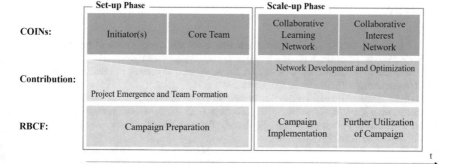

Fig. 19.1 Integrated model combining RBCF and COINs

quite well in their contributions. Especially early in the set-up phase, innovative endeavors can benefit from the application of an RBCF campaign to support the project emergence and to facilitate the team formation. In particular, the preparation of an RBCF campaign promotes the emergence of a project as it makes it easier for the initiators to further develop their early project idea during the preparation of the RBCF campaign as a specific sub-project. On the one hand, the campaign preparation provides a simplified and experimental setting to test what works for the project as well as for the team and what does not. On the other hand, such a "sandbox" project also facilitates the team at an early stage in terms of forming a shared social identity as well as developing roles, routines, and coordination processes within the team.

The further the team formation has progressed and the more the networks with external stakeholders have developed, the more does an innovation endeavor benefit of the application of the COINs approach. Therefore, the contributions of the COINs approach increase during the preparation, implementation, and further utilization of an RBCF campaign. Especially, in the scale-up phase the network metrics and conceptual tools of the COINs framework help to further develop and optimize the internal and external networks of the core team. For instance, honest signals and virtual mirroring could be applied to investigate social networks of RBCF teams and their communities to foster different kinds of campaign success as well as the further development of the innovation endeavor after the RBCF campaign has been implemented.

This is a conceptual paper that shows how RBCF campaigns complement the COINs approach to foster the emergence of innovative endeavors and their collaborative development. Future research might analyze empirically how RBCF campaigns should be designed to set up collaborative innovation networks for specific purposes. In this regard, it also would be interesting to investigate how several incremental RBCF campaigns could be applied to develop collaborative innovation networks in several steps over time. Furthermore, it might be analyzed in case studies how the COINs approach can be applied to optimize the network development and further utilization of social communities after an RBCF campaign has ended.

References

1. A. Agarwal, C. Catalini, A. Goldfarb, Some simple economics of crowdfunding. Soc. Sci. Res. Netw. (SSRN) Electron. J. (2013). http://ssrn.com/abstract=2281044
2. G.K. Ahlers, D. Cumming, C. Günther, D. Schweizer, Signaling in equity crowdfunding. Entrepreneurship Theor. Pract. 39(4), 955–980 (2015)
3. M. Beier, Startups' experimental development of digital marketing activities. A case of online-videos. Soc. Sci. Res. Network (SSRN) Electron. J. (2016). http://ssrn.com/abstract=2868449
4. M. Beier, Die Entwicklung Sozialer Netzwerke von Gründerteams, Dissertation thesis. Universität zu Köln, 2011
5. M. Beier, S. Früh, C. Jäger, Reward-based crowdfunding as a marketing tool for established smes: a multi case study. Soc. Sci. Res. Netw. (SSRN) Electron. J. (2019). http://ssrn.com/abstract=3338084

6. M. Beier, S. Früh, K. Wagner, Crowdfunding für unternehmen—plattformen, projekte und erfolgsfaktoren in der schweiz. Soc. Sci. Res. Netw. (SSRN) Electron. J. http://ssrn.com/abstract=2430147

7. M. Beier, K. Wagner, What determines the growth expectations of early-stage entrepreneurs? Evidence from crowdfunding. Int. J. Entrepreneurship Small Bus. 31(1), 12–31 (2017)

8. M. Beier, Wagner, *User Behavior in Crowdfunding Platforms—Exploratory Evidence from Switzerland*, in Proceedings of the 49th Hawaii International Conference on System Sciences (HICSS). Kauai, Hawaii, USA, 2016)

9. M. Beier, K. Wagner, *Crowdfunding Success: A Perspective from Social Media and E-Commerce*, in Proceedings of the 36th International Conference on Information Systems (ICIS). (Fort Worth, Texas, USA, 2015)

10. P. Belleflamme, T. Lambert, A. Schwienbacher, Crowdfunding: tapping the right crowd. J. Bus. Ventur. 29(5), 585–609 (2014)

11. P. Belleflamme, N. Omrani, M. Peitz, The economics of crowdfunding platforms. Inf. Econ. Policy 33, 11–28 (2015)

12. S. Bi, Z. Liu, K. Usman, The influence of online information on investing decisions of reward-based crowdfunding. J. Bus. Res. 71, 10–18 (2017)

13. S. Blank, Why the lean start-up changes everything. Harvard Bus. Rev. 91(5), 63–72 (2013)

14. T.E. Brown, E. Boon, L.F. Pitt, Seeking funding in order to sell: crowdfunding as a marketing tool. Bus. Horiz. 60(2), 189–195 (2017)

15. B. Chollet, S. Brion, V. Chauvet, C. Mothe, M. Géraudel, NPD projects in search of top management support: the role of team leader social capital. Management 15(1), 44–75, (2012)

16. M.G. Colombo, C. Franzoni, C. Rossi-Lamastra, Internal social capital and the attraction of early contributions in crowdfunding. Entrepreneurship Theor. Pract. 39(1), 75–100 (2015)

17. D.J. Cumming, G. Leboeuf, A. Schwienbacher, Crowdfunding models: keep-it-all versus all-or-nothing. Financ. Manag. Forthcoming

18. M. De Maggio, P.A. Gloor, G. Passiante, Collaborative innovation networks, virtual communities and geographical clustering. Int. J. Innov. Reg. Dev. 1(4), 387–404 (2009)

19. N. Eiteneyer, D. Bendig, M. Brettel, Social capital and the digital crowd: involving backers to promote new product innovativeness. Res. Policy 48(8), 103744 (2019)

20. D. Frydrych, A.J. Bock, Bring the noize: syndicate and role-identity co-creation during crowdfunding. SAGE Open 8(4), 1–15 (2018)

21. S. Gao, Y. Guo, J. Chen, The performance of knowledge collaboration in virtual teams: an empirical study. Int. J. Multimedia and Ubiquitous Eng. 9(8), 193–212 (2014)

22. E.M. Gerber, J.S. Hui, P.Y. Kuo, *Crowdfunding: Why People are Motivated to Post and Fund Projects on Crowdfunding Platforms*, in Proceedings of the International Workshop on Design, Influence, and Social Technologies: Techniques, Impacts and Ethics (Evanston, Illinois, USA, 2012)

23. F. Giones, A. Brem, Crowdfunding as a Tool for Innovation Marketing: Technology Entrepreneurship Commercialization Strategies, in *Handbook of Research on Techno-Entrepreneurship, Ecosystems, Innovation and Development*, ed. by F. Thérin, F.P. Appio, H. Yoon (Edward Elgar Publishing, Cheltenham, UK, 2019), pp. 156–174

24. G. Giudici, M. Guerini, C. Rossi-Lamastra, Why crowdfunding projects can succeed: the role of proponents' individual and territorial social capital. Soc. Sci. Res. Netw. (SSRN) Electron. J. (2013). http://ssrn.com/abstract=2255944

25. P.A. Gloor, *Swarm Creativity: Competitive Advantage through Collaborative Innovation Networks* (Oxford University Press, New York, USA, 2006)

26. P.A. Gloor, *Sociometrics and Human Relationships: Analyzing Social Networks to Manage Brands, Predict Trends, and Improve Organizational Performance* (Emerald Publishing Limited, Bingley, UK, 2017)

27. P.A. Gloor, *Swarm Leadership and the Collective Mind: Using Collaborative Innovation Networks to Build a Better Business* (Emerald Publishing Limited, Bingley, UK, 2017)

28. P. Gloor, A. Fronzetti Colladon, G. Giacomelli, T. Saran, F. Grippa, The impact of virtual mirroring on customer satisfaction. J. Bus. Res. 75, 67–76 (2017)

29. P.A. Gloor, R. Laubacher, S.B. Dynes, Y. Zhao, *Visualization of Communication Patterns in Collaborative Innovation Networks-Analysis of some W3C Working Groups,* in Proceedings of the Twelfth International Conference on Information and Knowledge Management (CIKM). (New Orleans, Louisiana, USA, 2003)

30. M.D. Greenberg, E.M. Gerber, *Learning to Fail: Experiencing Public Failure Online through Crowdfunding,* in Proceedings of the SIGCHI Conference on Human Factors in Computing Systems. (Toronto, Canada, 2014)

31. P. Haas, I. Blohm, J.M. Leimeister, *An Empirical Taxonomy of Crowdfunding Intermediaries,* in Proceedings of the 22th European Conference on Information Systems (ECIS). (Tel Aviv, Israel, 2014)

32. F. Hervé, A. Schwienbacher, Crowdfunding and innovation. J. Econ. Surv. **32**(5), 1514–1530 (2018)

33. C. Hopp, J. Kaminski, F. Piller, accentuating lead user entrepreneur characteristics in crowdfunding campaigns. The role of personal affection and the capitalization of positive events. J. Bus. Ventur. Insights **11**, e00106 2019

34. Y. Kim, A. Shaw, H. Zhang, E. Gerber, *Understanding Trust amid Delays in Crowdfunding,* in Proceedings of the 20th ACM Conference on Computer Supported Cooperative Work and Social Computing (CSCW). (Portland, Oregon, USA, 2017)

35. J.A. Koch, M. Siering, *Crowdfunding Success Factors: The Characteristics of Successfully Funded Projects on Crowdfunding Platforms,* in Proceedings of the 37th European Conference on Information Systems (ECIS). (Münster, Germany, 2015)

36. S. Kraus, C. Richter, A. Brem, C.F. Cheng, M.L. Chang, Strategies for reward-based crowdfunding campaigns. J. Innovation Knowl. **1**(1), 13–23 (2016)

37. V. Kuppuswamy, B.L. Bayus, in *A Review of Crowdfunding Research and Findings,* ed. by P.N. Golder, D. Mitra. Handbook of Research on New Product Development, (Cheltenham, UK: Edward Elgar Publishing, 2018), pp. 361–373

38. L.S.L Lai, E. Turban, Groups formation and operations in the web 2.0 environment and social networks. Group Decis. Negot. **17**(5), 387–402 (2008)

39. D. Leone, F. Schiavone, Innovation and knowledge sharing in crowdfunding: how social dynamics affect project success. Technol. Anal. Strateg. Manag. **31**(7), 803–816 (2019)

40. E. Mollick, The dynamics of crowdfunding: an exploratory study. J. Bus. Ventur. **29**(1), 1–16 (2014)

41. E. Mollick, V. Kuppuswamy, After the campaign: outcomes of crowdfunding. Soc. Sci. Res. Netw. (SSRN) Electron. J. (2014). http://ssrn.com/abstract=2376997

42. A. Moritz, J.H. Block, Crowdfunding: A Literature Review and Research Directions, in *Crowdfunding in Europe,* ed. by D. Brüntje, O. Gajda (Springer, Cham, Switzerland, 2016), pp. 25–53

43. O. Perdikaki, S. Kesavan, J.M. Swaminathan, Effect of traffic on sales and conversion rates of retail stores. Manufact. Serv. Oper. Manag. **14**(1), 145–162 (2012)

44. F. Piller, P. Schubert, M. Koch, K. Möslein, Overcoming mass confusion: collaborative customer co-design in online communities. J. Comput. Mediated Commun. **10**(4), JCMC1042 (2005)

45. O. Posegga, M. P. Zylka, K. Fischbach, *Collective Dynamics of Crowdfunding Networks,* in Proceedings of the 48th Hawaii International Conference on System Sciences (HICSS). (Kauai, Hawaii, USA, 2015)

46. S. Ryu, Y.G. Kim, A typology of crowdfunding sponsors: birds of a feather flock together? Electron. Commer. Res. Appl. **16**, 43–54 (2016)

47. A. Schwienbacher, B. Larralde, Crowdfunding of Small Entrepreneurial Ventures, in *The Oxford Handbook of Entrepreneurial Finance,* ed. by D. Cumming (Oxford University Press, New York, USA, 2012), pp. 369–391

48. T. Semrau, M. Beier, How specialised and integrated relationship management responsibilities foster new ventures' network development. Int. J. Entrepreneurial Ventur. **7**(1), 47–64 (2015)

49. Y. Song, R. Berger, Relation between start-ups' online social media presence and fundraising. J. Sci. Technol. Policy Manag. **8**(2), 161–180 (2017)

50. M.A. Stanko, D.H. Henard, Toward a better understanding of crowdfunding, openness and the consequences for innovation. Res. Policy **46**(4), 784–798 (2017)
51. M.A. Stanko, D.H. Henard, How crowdfunding influences innovation. MIT Sloan Manag. Rev. **57**(3), 15–17 (2016)
52. F. Thies, A. Huber, C. Bock, A. Benlian, S. Kraus, Following the crowd—does crowdfunding affect venture capitalists' selection of entrepreneurial ventures? J. Small Bus. Manage. **57**(4), 1378–1398 (2019)
53. F. Thies, M. Wessel, A. Benlian, *Understanding the Dynamic Interplay of Social Buzz and Contribution Behavior within and between Online Platforms—Evidence from Crowdfunding,* in Proceedings of the 35th International Conference on Information Systems (ICIS). (Auckland, New Zealand, 2014)
54. B.W. Tuckman, M.A.C. Jensen, Stages of small-group development revisited. Group Organ. Stud. **2**(4), 419–427 (1977)
55. J. Viotto da Cruz, Beyond financing: crowdfunding as an informational mechanism. J. Bus. Ventur. **33**(3), 371–393 (2018)
56. J. Wu, H. Sun, Y. Tan, Social media research: a review. J. Syst. Sci. Syst. Eng. **22**(3), 257–282 (2013)
57. Xu, X. Yang, H. Rao, W.T. Fu, S. W. Huang, B.P. Bailey, *Show Me the Money! An Analysis of Project Updates during Crowdfunding Campaigns,* in Proceedings of the SIGCHI Conference on Human Factors in Computing Systems. (Toronto, Canada, 2014)

Chapter 20
An Ecosystem for Collaborative Pattern Language Acquisition

Yuki Kawabe and Takashi Iba

Abstract In this paper, we describe an ecosystem for acquiring pattern languages from the perspectives of constructivism and collaborative way and introduce a web system called "Presen Box" that assists in pattern languages acquisition according to the ecosystem. Pattern language is a methodology that describes practical knowledge for enhancing human creativity and has been developed in various fields such as architecture, software design, education, organization, and lifestyle. In recent years, interfaces of pattern languages, pattern cards, pattern apps, and pattern objects that embed patterns in daily life, etc., have been developed in addition to reading materials such as books and papers. However, there are still things that need to be overcome in order for those who do not know of pattern languages to acquire it and promote higher quality practices. Rather than leaving pattern language acquiring to individual efforts alone, we propose an ecosystem that realizes collaborative acquisition and an "action first pattern practicing method" that supports the ecosystem. In this method, people will learn patterns through the repetition of concrete actions. The system "Presen Box" that implemented this ecosystem is a web platform that uses presentation patterns that describe presentation skills, and the users can get ideas for creating high-quality presentations. By repeating the execution of ideas, the users can acquire presentation patterns gradually.

Introduction

Pattern Language is a tool to enhance creativity in individuals and organizations. It is used as a common language and building blocks of thinking by verbalizing closed knowledge in an individual. Adam Grant, an organizational psychologist says, "Organizations that people help each other, share knowledge and care for each other are

Y. Kawabe (✉)
Faculty of Environment and Information Studies, Keio University, Tokyo, Japan
e-mail: t16279yk@sfc.keio.ac.jp

T. Iba
Faculty of Policy Management, Keio University, Tokyo, Japan

© Springer Nature Switzerland AG 2020
A. Przegalinska et al. (eds.), *Digital Transformation of Collaboration*,
Springer Proceedings in Complexity,
https://doi.org/10.1007/978-3-030-48993-9_20

better with all measurable indicators: profitability, customer satisfaction, employee retention rate, even for operating cost reductions [1]", sharing knowledge has a positive impact on organizations in a variety of ways.

To spread pattern language as a more effective tool, we have to develop not only pattern language itself but also interfaces that connect human and pattern languages as well as methods that make it easy for anyone to use patterns. In this paper, we propose an ecosystem that allows more people to learn and acquire pattern languages.

We first provide an overview of the pattern language. Then, we propose the ecosystem with a pillar "Action first pattern practicing method" that supports the ecosystem. Finally, we describe "Presen Box"[1] as a concrete example of an application that implemented the ecosystem and use cases. "Presen Box" is a web system that supports the creation of high-quality presentation by using the presentation patterns, a pattern language that describes expert knowledge of presentations.

Pattern Language

Overview of Pattern Language

Pattern language is a method of structuring patterns that are created by verbalizing and abstracting experts' secret skills. Each pattern describes a solution to a problem that is likely to occur in a particular context and is given a name. We can share the knowledge by using this as a vocabulary of communication.

Pattern Language is used in various fields. It was invented by an architect named Christopher Alexander, who developed "A Pattern Language" [2], a book that contains 253 patterns for architecture design. This method was then applied to software design [3], and many pattern languages have been developed. Since then, patterns have begun to be developed in all fields, such as Fearless Change [4] and Pedagogical Pattern [5]. We also created pattern languages about human actions, such as the learning patterns [6], presentation patterns [7], collaboration patterns [8], a pattern language for living well with dementia [9], and project design patterns [10].

Pattern Language Usage

There are three ways to use pattern languages, "Vocabulary for communication", "Glasses of Recognition", and "Triggers to generate new ideas" [11].

As "Vocabulary for communication", it is used as vocabulary, and people can share their knowledge when the skills have names. For example, there is the dialogue workshop. In this workshop, participants talk with each other about past episodes

---

[1] https://presentation.patternapp.net.

related to a particular pattern and get a practical image of the pattern and discover new usage of patterns. It is hard to talk about past learning experiences without topic, for example, the learning patterns will be a topic. What participants get from the workshop is useful when using the pattern in the future.

As "Glasses of Recognition", it is used when evaluating other's actions and looking back on their own action in a particular domain. For example, when they are looking back on their action, they evaluate the practice of patterns and recognize patterns that should be incorporated, after that they will be conscious of practicing the pattern when they take the action again.

As "Triggers to generate new ideas", it is used as a way to generate new ideas by getting perspective from a specific pattern. For example, there is an idea workshop using pattern cards as materials for ideas. In this workshop, the participants gain a new perspective by considering what kind of services should be available to realize the chosen pattern or what they can do if the chosen pattern is incorporated into current projects.

Patterns are also used in various ways not mentioned above. However, in order to actually use patterns in daily life, it is necessary to first remember the pattern users want to use and then recall it at an appropriate timing [12]. Various tools and methods, like the dialogue workshop [13], pattern cards [11], pattern objects [12], and pattern apps [14] are developed to make it easy to use patterns and to image how to practice them. However, there are still things to overcome for recalling patterns. Users must first meta-recognize the situation in which they are placed, which is difficult to do if they are too focused on only what is right in front of them. What is a better way to use patterns? The solution we propose is the process of acquiring patterns while repeating concrete actions. In order to realize this process, the collaboration between people who are able to practice patterns and those who are going to practice it and the ecosystem of practicing patterns will be important. Rather than leaving pattern practice to individual efforts alone, we think that more people can practice by practicing through collaboration among multiple people.

Ecosystem of Pattern Acquisition

Patterns are abstract concepts created from several experts' skills and used by users as a cognitive structure. The pedologist David Kolb proposes an experiential learning model and says that skills can be learned by repeating concrete experiences and being abstractly conceptualized through reflective observation and refined with active experimentation [15]. Additionally, the Swiss psychologist Jean Piaget says that human cognitive structure is constructed in individuals [16]. As shown in Fig. 20.1a, people don't construct cognitive structure by getting and learning from outside as taught by a teacher at school, but by experiencing and getting feedbacks from the surrounding environment through their own actions as shown in Fig. 20.1b.

As seen in Fig. 20.2, pattern practice has been done in the process of knowing patterns from books or websites, then recalling the pattern in an appropriate timing

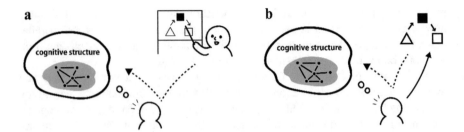

Fig. 20.1 **a** Learning cognitive structure from outside, **b** Developing cognitive structure through one's own action

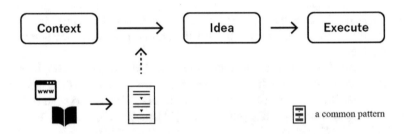

Fig. 20.2 Pattern practicing process when patterns are learned from outside

and generating a concrete idea by using it as "triggers to generate new ideas" and executing it.

However, the real situation changes every second, and there is no time to even recall it for people who don't acquire patterns. For example, right before the presentation or when you are giving it.

Based on Piaget's thesis, the patterns literally given and remembered cannot be used in the real situation. The patterns constructed while repeating specific actions are the ones that can be used. Alexander also said "It is not enough to merely duplicate a pattern from a book in order for each human being to keep in mind the pattern language as an expression of their daily life attitude" and "A language is a living language only when each person in society, or in the town, has his own vision of this language" [17], in order to practice patterns in a true sense, instead of imitating the written pattern as it is, people need to learn it as their own personalized model.

As shown in Fig. 20.3, beginners need to learn the pattern by repeating the process of adopting a specific idea that can be used in a specific context and executing it immediately again and again. We call this learning method "Action First Pattern Practice".

Figure 20.4 shows the practice process of people, who have personalized patterns. Instead of patterns given from the outside, personalized patterns can be extracted immediately in the real situation. Then, it will be used as "triggers to generate new ideas" to generate ideas that are suitable for the situation.

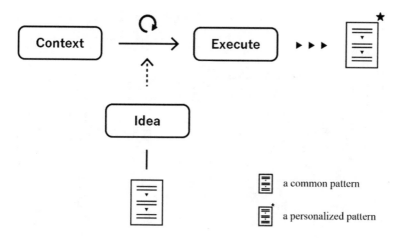

Fig. 20.3 Action first pattern practicing: Learning patterns through repetition of idea execution

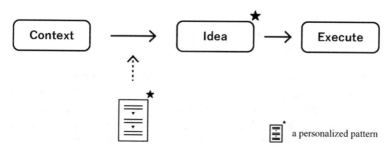

Fig. 20.4 Process of practicing patterns for people who have personalized patterns

The important thing about this pattern acquisition method is that the idea must be tied with the pattern. Collecting ideas that are tied with patterns is not easy for those who have not yet learned the patterns. Therefore, we solve it by showing the ideas tied to patterns of people who have already acquired the pattern to the person who is going to learn it. At this time, the pattern is used as "Vocabulary of communication" and "Glasses of Recognition".

Figure 20.5 shows the ecosystem that supports pattern learning. Phase 1 shows the learning process of the person who is going to learn, while Phase 2 shows the practice process of the person who has already learned. The idea to be introduced in Phase 1 is the idea that someone generated in Phase 2 in the past. Beginners execute the ideas that seniors have generated. Personalized patterns are not shareable, but they all link to patterns as a common language. The generated ideas are shared through patterns as a common language.

In addition, people who have learned it can refine their pattern practice by looking at other people's ideas.

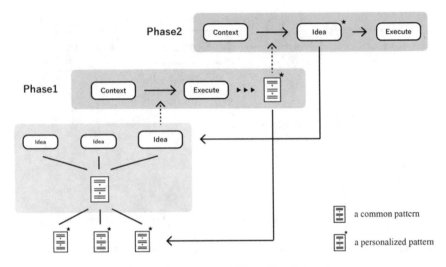

Fig. 20.5 An ecosystem of pattern language acquisition with collaboration

Implemented System: Presen Box

"Presen Box" is a web system that shares ideas for creating a better presentation using Presentation Patterns that describe good behaviors for presentations. The ideas for the presentation are lined up, and the user looks at the ideas that they may like. Users will access this in all situations and get hints when creating materials for the presentation, when considering the composition of the presentation, when trying to relieve tension just before the performance, and when looking back over a past presentation.

In the listing page (Fig. 20.6a), users choose ideas they like from listed ideas, and if there is something they can do, execute it. Ideas are listed with a card design and have titles, thumbnails, references, and patterns that are tied to the idea (Fig. 20.6b).

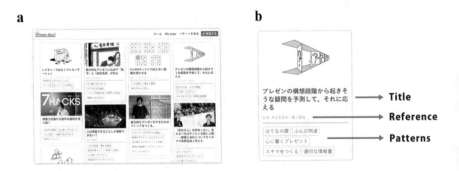

Fig. 20.6 a Idea listing page, **b** Card design of an idea

Each idea is tied with patterns and on the idea detail page (Fig. 20.7) other ideas that are tied with the same patterns are displayed as a recommendation.

The pattern detail page (Fig. 20.8) shows the description of the pattern and the ideas tied with it. As described above, users can search for new ideas in three ways, selecting from the list, choosing another idea from idea through tied pattern, or choosing an idea from a specific pattern.

The feature of the way of selecting from the list is to enable users who do not know about pattern languages to try patterns first and solve the problem that users cannot be put into practice without learning the pattern. All they have to do is choose and practice presentation ideas even if they do not know the patterns.

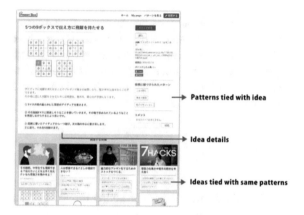

Fig. 20.7 Idea detail page

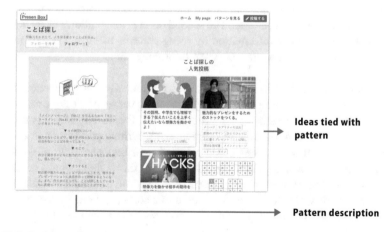

Fig. 20.8 Pattern detail page

Choosing and practicing another idea with the same pattern from one idea helps to understand the pattern. By repeating the practice of multiple ideas that are tied to the same pattern, they will gradually understand what the pattern means.

Tying patterns with a particular pattern will increase the searchability. For example, if users are nervous just before the presentation, they can ease the tension by acting the ideas that could be found from the pattern details page of [Best Effort] and [Construction of Confidence]. In addition, when preparing to make a presentation that attracts the audience, they can refer to the ideas linked to [Dramatic Modulation], [Doors of Mystery], and [Significant Void].

Users can follow the patterns. Ideas tied with following patterns are displayed preferentially on the list. You can stock your favorite ideas and find out immediately when you use them in the future.

When posting an idea, users input the title, the content, an image, and the pattern associated with it. It is difficult at first to post ideas for those who don't know the pattern so much, but it will be possible with repeating practice and getting a deep understanding of the pattern. The process of recognizing abstract concepts from concrete actions and generating specific ideas from abstract concepts further deepens the understanding of pattern languages.

Use Cases of Presen Box

In this section, we introduce three important points found out from the interview to the user group and explain how the "Presen Box" contributed to the acquisition and practice of the patterns on its ecosystem. The interviewee knew the presentation patterns and already practiced some of them, but the other patterns had not yet been mastered. The first one pointed out is that the "Presen Box" promotes the users to link actual presentations to patterns. The second point is that it allows users to search and collect higher quality information by using patterns as tags. The last point is that it can be used as a support tool when improving someone's presentation with patterns.

Good presentation practices can be used as reference whenever needed, if they are well organized. "Presen Box" helps users organize the set of ideas extracted from good presentation practices using patterns, and thus, it naturally reminds the users to link the practices with patterns. In addition, while writing paragraphs to share the important ideas with others, they could gain a deeper understanding of what each pattern actually means and what the good practice in this context would be like.

Since all the ideas posted on the "Presen Box" needs to be somehow linked to the patterns, the quality of information is ensured to a certain extent. By using patterns as search tags, users can quickly reach higher quality articles and practical ideas that meet a specific purpose, compared to when searching with search engines. In fact, a lot of articles are written only for gaining the number of accesses based on the SEO algorithm. The "Presen Box", however, provides users a platform to find high-quality information as fast as possible.

When commenting on someone's presentation using patterns, it would be much easier if you have some concrete examples. Even if the instructors have no such concrete example, the "Presen Box" shows some which would help learners understand the suggestion most effectively. Furthermore, it is very convenient in a way that it enables an online lecture by sharing a link.

As described above, it was found out that there are several advantages in enhancing one's learning and practicing patterns and sharing information in a small group. In the future, we would like to reflect on what happens if this tool is used by people who don't know about the pattern language at all and if it is introduced in a community in which many people belong to.

Conclusions

In this paper, we proposed an ecosystem that enhances pattern practicing and acquisition and "Action first pattern practicing" method that supports the ecosystem and presented "Presen Box" which is a web system based on them. The method of recognizing patterns by repeating concrete actions promotes the practice for those who did not learn the pattern so far and contributes to a deep understanding of the pattern for those who already learned the pattern. However, in order to share knowledge and improve collaboration using pattern languages, there is a need for further exploration of how individuals can get involved in the process. For example, "Presen Box" has been implemented as a web system but connecting pattern objects and web system will enable smoother practice by linking the Internet with the real world. We would like to test the feasibility of this system by using it in daily situations and collaborative works and improve effectiveness.

References

1. A. Grant, *Give and Take: Why Helping Others Drives Our Success* (Weidenfeld & Nicolson, London, 2013)
2. C. Alexander, S. Ishikawa, M. Silverstein, M. Jacobson, I. Fiksdahl-King, S. Angel, *A Pattern Language: Towns, Buildings* (Oxford University Press, NY, Construction, 1977)
3. K. Beck, W. Cunningham, Using Pattern Languages for Object-oriented Programs, in *OOPSLA-87 Workshop on the Specification and Design for Object-Oriented Programming* (Orlando, FL, 1987)
4. M.L. Manns, L. Rising, *Fearless Change: Patterns for Introducing New Ideas* (Addison-Wesley Professional, Boston, MA, 2004)
5. Pedagogical Patterns Editorial Board, *Pedagogical Patterns: Advice for Educators* (Joseph Bergin Software Tools, San Bernardino, CA, 2012)
6. T. Iba, Iba Laboratory, *Learning Patterns: A Pattern Language for Creative Learning* (CreativeShift Lab, Yokohama, 2014)
7. T. Iba, Iba Laboratory, *Presentation Patterns: A Pattern Language for Creative Presentations* (CreativeShift Lab, Yokohama, 2014)

8. T. Iba, Iba Laboratory, *Collaboration Patterns: A Pattern Language for Creative Collaborations* (CreativeShift Lab, Yokohama, 2014)
9. T. Iba, M. Okada (eds.), *Iba Laboratory, Dementia Friendly Japan Initiative,* in Words for a Journey: The Art of Being with Dementia (CreativeShift Lab, Yokohama, 2015)
10. T. Iba, F. Kajiwara (eds.), *Iba Laboratory, UDS Ltd.*, in Project Design Pattern (Shoeisha, Tokyo, 2016)
11. T. Iba, Pattern language 3.0; human action, dialog; workshop; behavioral properties, in *Pursuit of Pattern Languages for Societal Change,* ed by P. Baumgartner, T. Gruber-Muecke, R. Sickinger (Krems, 2016), pp. 200–233
12. T. Iba, A. Yoshikawa, T. Kaneko, N. Kimura, T. Kubota, Pattern Objects: Making Patterns Visible in Daily life, in *Designing Networks for Innovation and Improvisation Proceedings of the 6th International COINs Conference*, ed by M. Zylka, H. Fuehres, A. Colladon, P. Gloor (Switzerland, 2016), pp. 105–112
13. T. Iba, Pattern Language as Media for Creative Dialogue: Functional Analysis of Dialogue Workshop, in *PURPLSOC: The Workshop 2014*, ed by P. Baumgartner, R. Sickinger (Krems, 2014) pp. 212–231
14. H. Mori, Y. Kawabe, T. Iba, Patterns we live by: pattern app as a platform to familiarize pattern languages, in *Paper presented at the 7th Asian Conference on Pattern Languages of Programs* (Waseda University, Tokyo, 1–2 Mar 2018)
15. D. Kolb, *Experiential Learning: Experience as the Source of Learning and Development*, 2nd edn. (Pearson FT Press, NJ, 2014)
16. J. Piaget, B. Inhelder, Mental Imagery in the Child: A Study of the development of imaginal representation, translated from the French by P.A. Routledge, K Kegan Paul (London, 1971)
17. C. Alexander, *The Timeless Way of Building* (Oxford University Press, NY, 1979)

Printed in the United States
by Baker & Taylor Publisher Services